Analog Circuits and Signal Processing

Series Editors

The *Analog Circuits and Signal Processing* book series, formerly known as the *Kluwer International Series in Engineering and Computer Science*, is a high level academic and professional series publishing research on the design and applications of analog integrated circuits and signal processing circuits and systems. Typically per year we publish between 5-15 research monographs, professional books, handbooks, and edited volumes with worldwide distribution to engineers, researchers, educators, and libraries. The book series promotes and expedites the dissemination of new research results and tutorial views in the analog field. There is an exciting and large volume of research activity in the field worldwide. Researchers are striving to bridge the gap between classical analog work and recent advances in very large scale integration (VLSI) technologies with improved analog capabilities. Analog VLSI has been recognized as a major technology for future information processing. Analog work is showing signs of dramatic changes with emphasis on interdisciplinary research efforts combining device/circuit/technology issues. Consequently, new design concepts, strategies and design tools are being unveiled. Topics of interest include: Analog Interface Circuits and Systems; Data converters; Active-RC, switched-capacitor and continuous-time integrated filters; Mixed analog/digital VLSI; Simulation and modeling, mixed-mode simulation; Analog nonlinear and computational circuits and signal processing; Analog Artificial Neural Networks/Artificial Intelligence; Current-mode Signal Processing; Computer-Aided Design (CAD) tools; Analog Design in emerging technologies (Scalable CMOS, BiCMOS, GaAs, heterojunction and floating gate technologies, etc.); Analog Design for Test; Integrated sensors and actuators; Analog Design Automation/Knowledge-based Systems; Analog VLSI cell libraries; Analog product development; RF Front ends, Wireless communications and Microwave Circuits; Analog behavioral modeling, Analog HDL.

Mark Daniel Alea • Georges Gielen

Chips That Feel

Fully-Integrated Bioinspired Electronic Skin Systems

Springer

Mark Daniel Alea
KU Leuven
Leuven, Belgium

Georges Gielen
KU Leuven
Heverlee, Belgium

ISSN 1872-082X ISSN 2197-1854 (electronic)
Analog Circuits and Signal Processing
ISBN 978-3-032-26583-8 ISBN 978-3-032-26584-5 (eBook)
https://doi.org/10.1007/978-3-032-26584-5

This Springer imprint is published by the registered company Springer Nature Switzerland AG
The registered company address is: Gewerbestrasse 11, 6330 Cham, Switzerland

To my family, and to my wife Sharmine, your support kept me going; coffee did the rest.

Preface

Inspired by human touch, electronic skin (*e*-skin) systems consist of sensors that measure parameters like normal and shear forces and that transmit this information to central processing units to process and interpret the tactile stimuli. E-skin solutions enable humanoid robots and prosthetic users to receive tactile feedback that provides valuable information throughout the course of object manipulation. Detecting slip at its earliest onset, before it develops into sustained sliding that can lead to object loss, is crucial for the safe manipulation of fragile objects. Similarly, measuring fine-grained contact parameters such as the texture helps determine object properties. The amazing capabilities envisioned in future e-skin systems are, not surprisingly, so commonly and intuitively performed with human tactile perception. Such biomimicry of human skin, however, is technologically quite challenging as the human somatosensory (skin) system is astonishingly complex.

This book presents research that bridges the gap between existing *e*-skin technologies and the remarkable capabilities of human tactile perception. It aims to realize a more human-like *e*-skin system that mirrors the essential properties of human skin: *multi-modal*, *multi-scale*, and capable of *local processing*. Transitioning from off-the-shelf solutions to a fully integrated implementation inherently yields system-level improvements. However, achieving human-level *e*-skin capabilities still demands a suite of complementary strategies and design solutions guided by a system-level, end-to-end perspective. More specifically, by incorporating neuromorphic design concepts, our goal is to advance the e-skin sensing spatial resolution, its power efficiency, the embedded processing, feature extraction, and data compression.

Building on this vision, we present end-to-end design strategies for next-generation, fully integrated e-skin systems. We begin by introducing a pragmatic, technology-aware neuromorphic design approach that draws inspiration from biology without blindly replicating it. One such aspect is the bioinspired encoding of tactile signals, showing the system-level benefits of event-based approaches compared to conventional frame-based methods due to its temporal compression of data (*i.e.,* via spikes). We also show how spike-based output representations naturally support efficient neuromorphic processing, by means of spiking neural networks (SNNs), which naturally relaxes the readout design constraints. This helps to realize the promise of true spatiotemporally sparse sampling (spatial and temporal compression), through the implementation of a per-taxel readout approach that is inspired from the same high-density and power-efficient sensing as in the human hands, especially the fingertips. We address the accompanying engineering challenges for realizing such per-taxel readout: chip area reduction, low-power constraints, and sensor miniaturization and integration, and we describe the solutions we have developed to address these challenges. Finally, we introduce our vision for a multi-modal, multi-scale e-skin architecture, by designing and validating the INTUITIVE and DERMIS works, which together form a pathway toward e-skin systems with human-like tactile perception and embedded intelligence.

Building on the above concepts, the INTUITIVE chip serves as a key implementation of our proposed end-to-end design strategy, incorporating spatiotemporally sparse sampling through per-taxel readout, enabled by the on-chip integration of tactile sensors on a dense CMOS readout chip in commercial technologies. The INTUITIVE chip, implemented in a 0.18-μm CMOS technology, achieves the highest reported sensing spatial resolution of 200 μm, comparable or even beyond human fingertips. A key innovation is the integration on chip of a 12×16 polyvinylidene fluoride (PVDF)-based piezoelectric sensor array with a per-taxel signal conditioning frontend and spiking readout combined with local embedded neuromorphic first-order processing through Complex Receptive Fields (CRFs). The experimental results confirm our hypothesis that the SNN-based classification of output spikes is inherently robust to the readout circuit's non-idealities, achieving high accuracies despite the area- and power-constrained per-taxel readout design. Finally, a comparison of the texture classification accuracies between two spike encoder outputs implemented on-chip shows that the proposed bioinspired neuromorphic level-crossing sampling (N-LCS) architecture with decaying threshold outperformed the conventional bipolar level-crossing sampling (LCS) architecture with fixed threshold, thus, proving the actual system-level benefits of our pragmatic technology-aware bioinspired approach.

The INTUITIVE chip is highly effective for fine, localized texture sensing, such as that performed by human fingertips. However, slip detection requires large-area sensing, which is cost-prohibitive for CMOS technologies, which therefore calls for implementation on a different substrate. Additionally, because the INTUITIVE system relies on piezoelectric sensors, it can only respond to dynamic normal forces (i.e., vibrations). In contrast, human skin exhibits multi-modal and multi-scale sensing capabilities, detecting both normal and shear forces—static and dynamic—across localized and extended regions alike. To extend the INTUITIVE chip toward large-area shear and normal force sensing, the DERMIS chip introduces a high-resolution, fully integrated tactile sensor system implemented in a flexible thin-film transistor (TFT) technology. By sensing both shear and normal forces, the DERMIS chip can inherently detect slip, mirroring the way human skin encodes slip through the spatiotemporal spike-based encoding of high-resolution shear and normal force signals during object manipulation. Furthermore, due to the analog-based encoding of these *grasp-state-dependent* contact parameters, the system avoids the need for complex offline slip-extraction algorithms. Similar to the INTUITIVE approach of implementing CRFs, embedded feature extraction is likewise implemented: to detect the various object properties throughout the course of grasp manipulation, embedded grasp-state-adaptive force extraction logic is implemented on chip.

Together, these two systems (INTUITIVE and DERMIS) establish a pathway toward multi-modal, multi-scale, and local-processing e-skin systems, combining high-resolution texture sensing with low-latency, large-area slip detection. In this way, they significantly advance beyond prior e-skin technologies, bringing electronic skin solutions closer to the capabilities of the human skin.

The authors acknowledge the support for this research from the KU Leuven C3 Project DERMIS and the European Union's Horizon 2020 research and innovation program under the Marie Skłodowska-Curie grant agreement No 861166 (INTUITIVE).

Leuven, Belgium
Heverlee, Belgium
March 2026

Mark Daniel Alea
Georges Gielen

Declarations

Competing Interests The authors have no competing interests to declare that are relevant to the content of this book.

Contents

Acronyms

a-IGZO	Amorphous indium-gallium-zinc oxide
AC	Alternating-current
ADC	Analog-to-digital converter
AdEx	Adaptive exponential
ADSM	Asynchronous delta-sigma modulator
AER	Address event representation
AFE	Analog front-end
AI	Artificial intelligence
ANN	Artificial neural network
ASIC	Application-specific integrated circuit
ATP	Adenosine triphosphate
AZ	Auto-zero
CAGR	Compound annual growth rate
CDC	Capacitance-to-digital converter
CDM	Code division multiplexing
CMFB	Common-mode feedback
CMOS	Complementary metal-oxide-semiconductor
CMP	Chemical-mechanical polishing
CNN	Convolutional neural network
COTS	Commercial off-the-shelf
CPU	Central processing unit
CRF	Complex receptive field
CS	Common-source
CT	Continuous-time
DAC	Digital-to-analog converter
DC	Direct current
DERMIS	Dense Electronic Readout for Multi-scale Integrated e-Skins (KUL project name)
DVS	Dynamic vision sensors
ECG	Electrcardiography
EEG	Electroencephalogram

ELM	Extreme learning machine
EMG	Electromyography
ENOB	Effective number of bits
ESD	Electrostatic discharge
FA	Fast-adapting
FFC	Flat flexible cable
FOM	Figure of merit
FSR	Force-sensitive resistor
GBW	Gain bandwidth
IF	Integrate-and-Fire
INTUITIVE	INnovative network for Training in toUch InteracTIVE interfaces (EU project name)
ISI	Inter-spike interval
LC	Level crossing
LC-ADC	Level-crossing analog-to-digital converter
LCS	Level-crossing sampling
LDO	Low-dropout regulator
LIF	Leaky integrate-and-fire
LSB	Least significant bit
LTPS	Low-temperature polycrystalline silicon
MCU	Microcontroller unit
MEMS	Micro- electro-mechanical system
ML	Machine learning
MOS	Metal-oxide-semiconductor
MOSFET	Metal oxide semiconductor field effect transistor
N-LCS	Neuromorphic level-crossing sampling
NLC-ADC	Neuromorphic level-crossing analog-to-digital converter
NMOS	N-type metal-oxide-semiconductor
NPU	Neural processing unit
NW	Nanowire
OTA	Operational transconductance amplifier
PCB	Printed circuit board
PDMS	Polydimethylsiloxane
PMOS	P-channel metal-oxide-semiconductor
POSFET	Piezoelectric oxide semiconductor field effect transistor
PSD	Power spectral density
PTAT	Proportional to absolute temperature
PVDF	Polyvinylidene fluoride
PVDF-TrFE	Polyvinylidene fluoride – trifluoroethylene barium titanate
RISC	Reduced instruction set computer
RMS	Root mean square
RNN	Recurrent neural network
SA	Slow-adapting
SAR	Successive approximation register
SEM	Scanning electron microscope

SER	Signal-to-reconstruction error ratio
SFDR	Spurious-free dynamic range
SMU	Source measurement unit
SNDR	Signal-to-noise-and-distortion ratio
SNN	Spiking neural network
SNR	Signal-to-noise ratio
SRM	Simplified response model
STDP	Spike-timing-dependent plasticity
SVM	Support vector machines
TFT	Thin-film technology
THD	Total harmonic distortion
VCO	Voltage-controlled oscillator
VLSI	Very large-scale integration
ZnO	Zinc oxide

Chapter 1
Introduction

Abstract This chapter starts by describing how the convergence of bioinspired electronic designs and the proliferation of smart, low-cost sensors usher in a new wave of promising biomedical devices with the potential to save and improve lives. Neuroprosthetic users seek to regain human-like touch—restoring functionality, environmental connection, and quality of life. Likewise, advanced robotics greatly benefit from high-fidelity tactile sensing to achieve more human-like grasping and manipulation. Then, artificial analogs of biological skin—electronic skins—are described and the challenges involved in achieving the performance and capabilities of their biological counterparts are described.After outlining these key challenges, a brief detour is taken into neuroscience to examine the intricacies of the human somatosensory system, aiming to identify fundamental principles that guide the development of human-like tactile perception in *e*-skin systems. There, the fundamental properties of human skins are highlighted: its multi-modal, multi-scale properties and its capability to perform local processing. These fundamental properties then form the guiding principle of this research, resulting in uniting the diverse sensing paradigms of the INTUITIVE and DERMIS projects into a cohesive, functional *e*-skin system.

Keywords Electronic skin · Bioinspired sensors · Tactile perception · Biomedical devices

1.1 Electronic Skin: Drawing on Human Touch Perception

The convergence of widespread biomimicry adoption and the growing availability of smart, low-cost sensors has not only made human life more comfortable, but has also ushered in a new wave of promising biomedical devices with the potential to save or improve lives. Artificial cochlear and retinal implants, both inspired by the inner workings of their biological counterparts, are already helping people who have lost their hearing or sight to regain vital sensory functions. Similarly, patients with prosthetic arms would like to regain human-like touch, enabling them to feel pressure, texture, and temperature in ways that restore not only that functionality

M. D. Alea, G. Gielen, *Chips That Feel*, Analog Circuits and Signal Processing,
https://doi.org/10.1007/978-3-032-26584-5_1

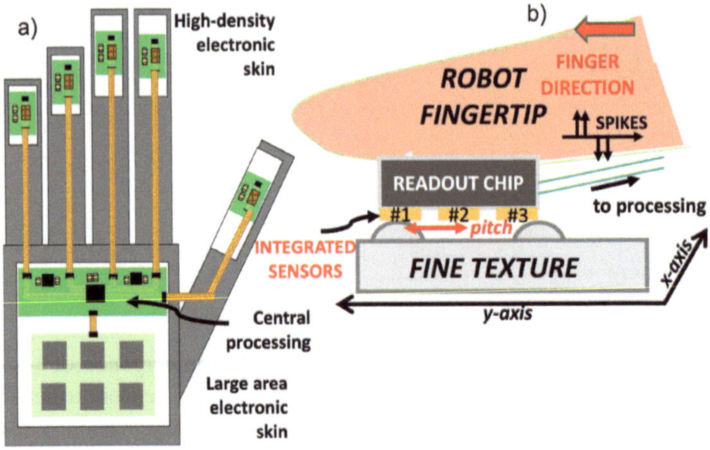

Fig. 1.1 (**a**) Equipping a robotic or a prosthetic hand with a combination of large-area e-skin in the palm and high-spatial-resolution e-skin on the fingertips allows the artificial hand to perform the classification of macro and micro tactile features. (**b**) High-spatial-resolution e-skin allow sensing the fine features of textured objects, hence allowing texture recognition

but also a sense of connection to their environment, and improving their quality of life.

In a like manner, the amount of robot applications in our society is rising, both in industrial as well as home applications (e.g. package delivery, person care, cleaning, etc.). Such robots would benefit immensely from improved human-like grasping capabilities, enabled by high-fidelity tactile sensation and feedback. Furthermore, in future industries, the frequent interaction between humans and robots underscores the critical importance of safety, supported by equipping robots with sensory capabilities.

To address both application fields, electronic skin ("e-skin") solutions are being developed that achieve human-like tactile capabilities (Bartolozzi et al., 2016; Dahiya et al., 2019) (Fig. 1.1).

E-skin has been called many names across literature. Among them is the more electronics-leaning ***tactile sensors***. A general observation for papers using this term is that they follow the lineage of prior works, where the implementation focus is on developing a precise readout for any general sensor technology (e.g., MEMS capacitive sensors for accelerometers, piezoresistive sensors for temperature sensing). That is, solving the tactile perception problem utilizes straightforward circuit design strategies with little or no bioinspired concepts. The term ***neuromorphic skins***, on the other hand, follows the research lineage of groups like in ETH Zurich (Indiveri et al.,

(continued)

2011), which incorporate biological concepts into the circuit implementations. Driven by neuroscience observations, the diffusion mechanism in a weak-inversion MOS device is, for instance, leveraged to emulate the brain's similar diffusion-based ion flow (Mead, 1989). What began as exclusively approximating the inner workings of the neurons in the brain has since expanded into drawing on the firing properties of afferents in the retina (Lichtsteiner et al., 2008), the skin (Janotte et al., 2022), and the like.

What is e-skin and what is it not? The similarity to other large-area sensing applications has understandably generated confusion as to what is an e-skin. For the sole purpose of this book, touch sensing for displays, and EMG/ECG patch-based sensing are not considered e-skin. For instance, touch sensing detects finger or stylus indentations, whereas tactile sensing/e-skin detects any general contact with an object without the need for prior tools (such as the aforementioned finger or stylus). Furthermore, other more exotic skin-inspired sensing such as artificial whiskers (Sullivan et al., 2012) (although an interesting and novel approach to tactile perception) is not considered to be e-skin in this book.

To mimic the touch in humans, e-skin systems consist of sensors that measure parameters like normal (Rosa & Kraft, 2024) and shear forces (Boutry et al., 2018), temperature (Yang et al., 2008), pain (Yoon et al., 2018), and proprioception (Dawood et al., 2020); and that transmit this information to central processing units (Rasouli et al., 2018) to interpret (Schmitz et al., 2019) and classify (Rongala et al., 2020; Virone et al., 2021; Canepa et al., 1998) the tactile stimuli. E-skins enable humanoid robots and prosthetic users to receive tactile feedback (Dahiya, 2019) that provides valuable information throughout the course of object manipulation. *Incipient* slip (partial slip), a precursor to *gross* slip (total loss of object contact) is a crucial parameter to detect for the safe manipulation of fragile objects (Romeo & Zollo, 2020). Similarly, measuring contact parameters such as the texture and hardness, and the surface temperature help determine the object's properties (Sundaram et al., 2019). The amazing capabilities envisioned in future e-skin systems are, unsurprisingly, so commonly and intuitively performed with human tactile perception. Such biomimicry of human skin, however, is technologically quite challenging as the human somatosensory (skin) system is astonishingly complex.

Human tactile sensation relies on sensing and encoding that involves:

- millisecond-precise (Mackevicius et al., 2012) asynchronous signal-to-spike tactile stimuli encoding (Pruszynski & Johansson, 2014; Saal & Bensmaia, 2014);
- complex and robust shear force encoding (Delhaye et al., 2021; Platkiewicz et al., 2016);
- local spatiotemporal (Lesniak et al., 2014; Pruszynski & Johansson, 2014; Saal et al., 2016) signal compression with dense receptor-afferent connections (Hay & Pruszynski, 2020);

- feature extraction that is dependent on the current grasp state during object manipulation (Willemet et al., 2021; Johansson & Flanagan, 2009); and
- fast (50–100 ms) in slip detection during grasping (Johansson & Westling, 1984).

We now look further at the intricacies of the human somatosensory system to find guiding principles that allow us to achieve the ultimate goal of electronically approximating or implementing human-level tactile perception.

1.2 Principles of the Human Skin: An Inspiration

The skin is the largest sensory organ. The first modern published work (Du Bois & Du Bois, 1916) calculating the body surface area yields estimates of around 1.7–1.9 m^2. Current estimates suggest young adults are innervated by ~230,000 tactile afferent fibers (Corniani & Saal, 2020). Primary afferent fibers are the very first step in the sensory pathway: they pick up signals from the environment or the body itself and carry them into the spinal cord and brain for processing (Fig. 1.2). Fifteen percent of the tactile afferent fibers are located in the hand regions, suggesting its importance in complex human tasks: grasping and manipulation. Evidence also suggests that local processing already happens at the primary afferents of the fingertips (Pruszynski & Johansson, 2014). Consequently, efforts to biomimic the sensing capabilities of the human skin at the fingertips and hand are likely to demand the most substantial technological advances.

In this section, we will focus mainly on the inner workings of the human somatosensory system in the hands, and more specifically, the fingertips.

1.2.1 Human Skins Are Multi-scale

The receptor density varies significantly (10–100×) (Corniani & Saal, 2020) at different areas across the entire human body, resulting in different sensing paradigms depending on the skin location. This property of the human skin is what we refer to as *multi-scale* (combines both local and distributed sensing; fine and coarse spatial resolution). Small localized regions in, for instance, the fingertips, used for fine texture and vibration sensing, have been reported to have sub-millimeter (mm) spatial resolution, while sensing in the hand palms, for instance, has a coarser density (> 1 mm) but covers a much larger area. The fingertips, while only occupying 0.3% of the human skin area, contains almost 5% of the tactile afferents (Corniani & Saal, 2020). This variation in the density of peripheral innervation is commonly depicted with the *homonculus*, wherein the relative scale of the body part signifies the innervation density (see Fig. 1.3). So, the required tactile capabilities vary widely in density and sensitivity along the arm/hand/finger.

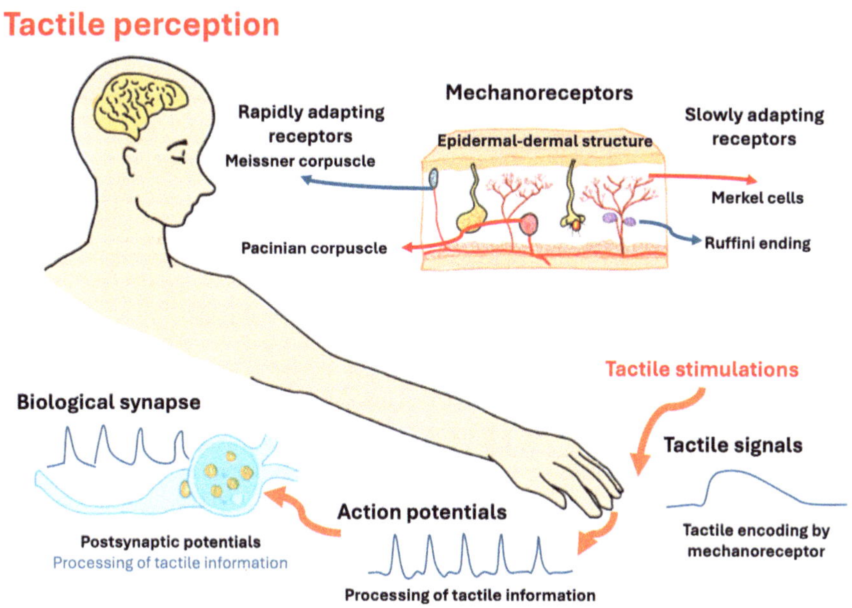

Fig. 1.2 Receptors of varying depth on the skin are tailored for different force profiles, be it static force, dynamic force, or vibration. Tactile signals are encoded into action potentials, which are processed locally then travel upstream to the spinal cord and to the brain for further processing (Adapted from Lee et al., 2019b)

Fig. 1.3 Tactile innervation densities for the whole body, as depicted with the *homonculus* (Reproduced from Corniani & Saal, 2020. Licensed under CC BY 4.0)

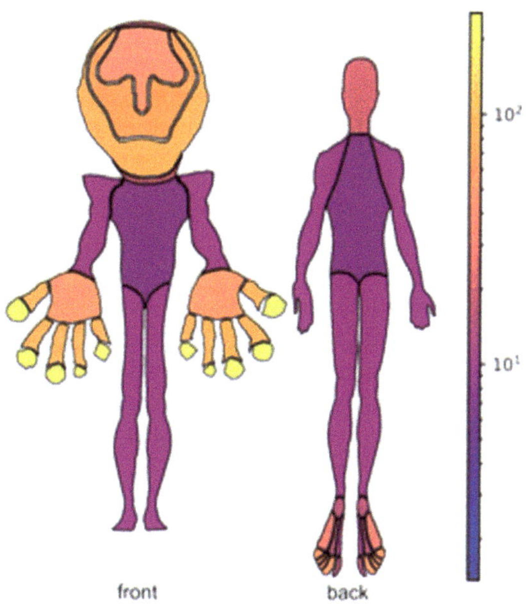

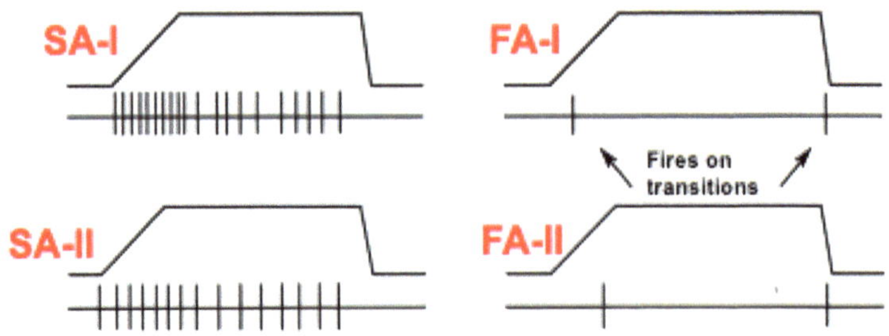

Fig. 1.4 Afferent firing behaviors of the four mechanoreceptors on the skin

Table 1.1 Cutaneous mechanoreceptors, adaptation rates, receptive fields, and sensitivities of the four mechanoreceptors on the human skin (Adapted from Skedung, 2012)

Mechanoreceptor	Adaptation rate	Receptive field	Sensitivity
Merkel disks	Slow (SA)	Small (I)	Pressure: deformation in spatial structure (2–16 Hz).
Ruffini end organ	Slow (SA)	Large (II)	Stretch: lateral deformation (100–500 Hz).
Meissner corpuscle	Fast (FA)	Small (I)	Flutter: light touch, movement, or deformation changes (2–40 Hz).
Pacinian corpuscle	Fast (FA)	Large (II)	Vibration: fine textures, movement, or deformation changes (40–500 Hz).

1.2.2 Human Skins Are Multi-modal

Human skins transduce a complex combination of tactile stimuli components, such as normal force, shear force, stretching, vibration, and temperature, using different specialized mechanoreceptors. This property of the human skin is what we refer to as *multi-modal* (combines receptors of varying sensitivity to force profiles and object properties). Specific mechanoreceptors embedded in the skin respond almost exclusively to either static or dynamic force, reinforcing the multi-modal nature. Skin receptors (Fig. 1.2) are divided into two major groups (Johansson & Vallbo, 1979): slow-adapting (SA) and fast-adapting (FA) afferents (Fig. 1.4). Merkel disks (SA-I) and Ruffini endings (SA-II) are sensitive to static contact events (i.e., the spike firing rates encode the magnitude of contact). Meanwhile, Meissner corpuscles (FA-I) and Pacinian corpuscles (FA-II) are sensitive to dynamic events (i.e., spikes at the forming and breaking of contact). Such specialization of the receptors corresponds to varying receptive field and force profile sensitivity, as summarized in Table 1.1.

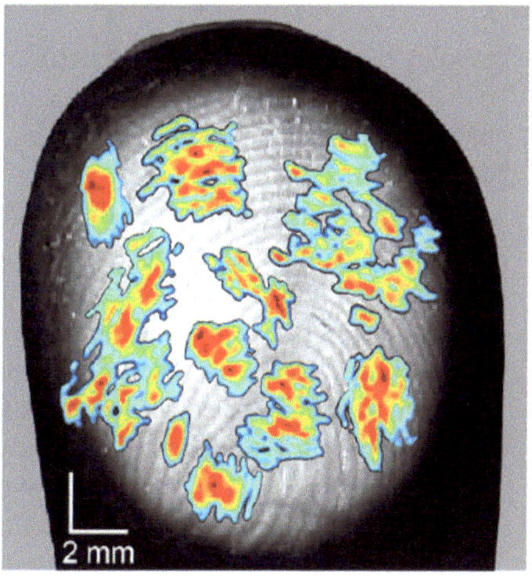

Fig. 1.5 Observed non-homogeneous complex receptive fields (CRFs) in the fingertips help to encode the tactile stimuli at the primary afferent level (Reproduced from Jarocka et al., 2021. Licensed under CC BY 4.0)

1.2.3 Human Skins Perform Local Processing

The multi-modal and multi-scale nature of human skins enable the somatosensory system to encode tactile stimuli with high fidelity. In fact, recent work (Pruszynski & Johansson, 2014; Jarocka et al., 2021) suggests that some form of tactile signal processing/compression occurs at the primary afferent level, much earlier in the somatosensory system pathway than previously assumed.

Human skin offers a form of spatial compression of sensor information through complex receptive fields (CRFs), as shown in Fig. 1.5. CRFs are formed by random innervation of multiple mechanoreceptors by a single afferent (Pruszynski & Johansson, 2014; Jarocka et al., 2021). Therefore multiple sensor information is compressed into a single spike train. It has been theorized that the CRFs allow for the first-order encoding of the tactile stimuli (Pruszynski & Johansson, 2014). This means that spike trains generated by the first-order neurons already reflect the tactile stimulus information such that the spike trains allow sufficient discriminability across different stimuli.

Several parallels can be inferred between groundbreaking work in *deep learning* architectures and *near-sensor computing*, and how the somatosensory processing system works. This is borne out of the observation that the human somatosensory neural pathway follows multilevel processing hierarchies. The CRFs at the primary afferent level can be viewed as some form of convolutional-network-based compression of spatial tactile features. That is, certain tactile stimuli such as edge orientation (Hay & Pruszynski, 2020) can be discriminated maximally by the mere formation

(continued)

topology of the CRFs at the fingertips. One layer upstream to the spinal cord, we see the *cuneate* nucleus, where the high-dimensional representation of haptic events is then extracted before passing it onwards to the brain (Jörntell et al., 2014).

Similarly, one can look at the processing at the primary afferent level to be similar to another emerging field in sensing: *edge* processing or *near-sensor* computing (Zhou & Chai, 2020; Safa et al., 2022). In contrast to the more hardware-intensive *cloud-based* processing, edge-based processing generally requires lower energy budgets and simpler hardware at the expense of less computational power, which is usually sufficient for simple tasks (i.e., compression, fast inference). In the case of its biological counterpart, primary afferents perform local processing early on, despite their relatively fewer numbers and limited synaptic connections (Lesniak et al., 2014), as compared to the incredible complexity of the brain (akin to cloud-based compute).

1.3 Analogues to Neuromorphic Vision

Until recently, electronic skin work has arguably not received the same mainstream attention as other biomedical applications, such as neural implants, EMG/ECG sensing, artificial cochlear implants, and neuromorphic vision. Therefore, when explaining fundamental requirements and implementation details for e-skins, one can point to its close relative as a tool: *neuromorphic vision*. Besides the obvious parallels between them as being two of the five senses that humans have, the two applications share similar sensing and processing challenges. The basic unit of a tactile sensor, called the *taxel* (tactile pixel), has obvious roots from image sensing. In addition, research in the two fields has benefitted from the intermingling between various disciplines, namely materials and sensor fabrication, very-large-scale integration (VLSI), neuroscience, and neuromorphic concepts (Chen et al., 2020).

Borrowing concepts from the inner workings of the retina (Fig. 1.6a), the pioneering work into dynamic vision sensors (DVS) was started by the group of Tobi Delbruck (Lichtsteiner et al., 2008). In contrast to standard frame-based cameras, DVS captures only the per-pixel brightness changes, delivering a lower power consumption and latency, and a higher dynamic range and temporal resolution. DVS, as compared to a standard frame-based camera, tracks very well fast-changing inputs (e.g., rotating dot on the circle periphery), while disregarding redundant data (e.g., stationary dot at the center) due to its event-based sampling. This spatiotemporal sensitivity to changes or *events* is inspired by the inner workings of ON- and OFF-bipolar cells, which code for the bright and dark spatiotemporal contrast changes. This spatiotemporal event-based sampling of the input is a concept that will be massively adopted and explored in this text.

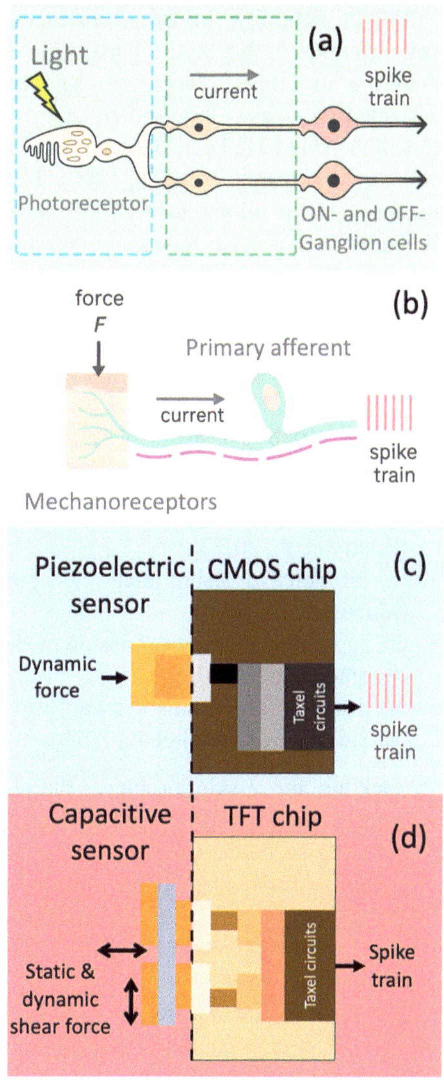

Fig. 1.6 (**a**) Light-to-spike conversion in biological retinas, (**b**) Force-to-spike conversion in biological skin, (**c**) Bioinspired dynamic force-to-spike conversion in e-skin, (**d**) Bioinspired multi-axial force-to-spike conversion in e-skin

The contribution of VLSI to the development of DVS (or image sensors, in general) should not be understated. Improvements in VLSI technology have concurrently enabled smaller sensors and electronics, and reducing the pitch of today's DVS sensors, now closely approximating the retinal capabilities. The 1:1 ratio between these two elements enabled true spatiotemporal sampling. Furthermore, the co-integration of the sensors and readout circuitry avoids long wires and bondpads, eliminating the related noise/bandwidth penalties (Tasker et al., 2024). As we will see later in Chap. 3, the same approach of leveraging VLSI has been one of the key components in our work to achieve human-level tactile sensing capabilities.

E-skin, naturally, has broad distinctions over neuromorphic vision sensors. A key difference is that tactile sensors are exposed to mechanical contact (Fig. 1.6b), therefore are prone to wear and tear. This significantly complicates sensor design and integration, as will be shown later in Chap. 3. Furthermore, the mechanoreceptors in the skin have been shown to have a ~70× larger pitch than its photoreceptor counterparts (Curcio & Allen, 1990). This relaxes the per-taxel e-skin readout area requirement, or allows to introduce further circuit complexity to perform novel functions such as CRF-based compression, as will be performed in this work.

1.4 Challenges in E-skin Designs

Until now, progress on artificial touch has been lagging significantly behind artificial vision. This is due to the many challenges to be addressed:

- the complex mechanical translation of biological skin into artificial tactile sensors (Boutry et al., 2018);
- the difficulty to design area- and power-efficient electronic readout for e- skin (Murray et al., 2020);
- the aggregation and transmission of the rich tactile information over large areas (Bergner et al., 2020);
- the design of algorithms and hardware for accurate tactile stimuli detection and classification (Rasouli et al., 2018).

Tackling the e-skin problem, therefore, requires a system-level end-to-end approach (Bartolozzi et al., 2016; Dahiya et al., 2019; Van Assche et al., 2023). Given these technical challenges, it is not surprising that electronic skin has not yet seen a widespread adoption in robotics and prosthetics (Romeo & Zollo, 2020). A recent report (Ciancio et al., 2016), for example, has highlighted the virtually non-existent prevalence of tactile feedback in commercial prostheses. Yet, the applications are enormous and growing.

The current state of the art in e-skin designs attempts to mimic biology in several aspects: (a) asynchronous spike encoding of the tactile stimuli (Janotte et al., 2022), (b) large-area sensing (Lee et al., 2019a), (c) normal and shear force encoding (Boutry et al., 2018), (d) receptive field feature extraction (Ali et al., 2021), and (e) spike-based classification of the tactile stimuli (Rongala et al., 2020). However, key limitations in the existing state-of-the-art designs are their (a) low spatial sensing resolution and (b) the lack of embedded computation for tactile stimuli processing. Biological skin sensing on the other hand demands a multi-scale spatial resolution and involves embedded feature extraction computation early in the tactile processing pathway (Jarocka et al., 2021), as discussed above.

The confluence of emerging neuromorphic tactile and vision sensors, interestingly, ushers in an increased interest in using multi-modal sensing during complex robotic tasks (Mao et al., 2024). Such an approach, however, is fundamentally different from the multi-modal nature of human skins. Tactile-vision multi-modal sensing combines force and light sensing, whereas tactile sensing is still largely force-based sensing. Therefore, multi-modal approaches exclusively for tactile perception are still quite limited in literature (Navaraj & Dahiya, 2019).

Similarly, multi-scale approaches are also quite limited in existing state-of-the-art work. For example, the limited spatial resolution of existing tactile sensors relegates most e-skin solutions to the large-area coverage of the arms and legs (Dahiya et al., 2019). Prior e-skin work approximating the dense and high-spatial-resolution sensing in the fingertips uses sensors with below human-level resolution (Navaraj & Dahiya, 2019).

Our research aims at overcoming these limitations with actual e-skin designs.

1.5 E-skin: An Emerging Commercial Application

Publicly available market information about the robotic sensor market shows a wide range of the market size for tactile sensors in 2024: from $1.97B to $16.4B. The market is expected to grow with a compound annual growth rate (CAGR) between 11.1 to 15.86% (Gujar & Das, 2025; Grand View Research, 2024; Munde, 2025; Mordor Intelligence, 2025), with an estimated value reaching $35.59B by 2031 (Pednekar, 2025).

The market for sensorized robotic arms is expected to grow in the coming years, driven by factors such as the increasing demand for automation and safety in manufacturing (XELA Robotics, 2025), the rising demand for robots in medical applications (Othman et al., 2022; Osborn et al., 2014; Jarumethitanont et al., 2024), the increasing sensorization of car seat surfaces (BMW, 2024; Canatu, 2018), the need to automate manual inspections by hand (GelSight, 2020; Bartlett, 2023; TacterionGmbH, 2023), the rise of humanoid robots (Clark, 2025; Holgado et al., 2019), etc.

Some key players in the market for robotic arms that can detect fine tactile details, include ABB, FANUC, KUKA, Yaskawa Electric and Universal Robots, among others. These companies offer a range of robotic arm solutions with varying degrees of tactile sensing capabilities, as well as software and integration services to help manufacturers implement these systems into their operations. The cost of these systems can vary widely depending on the level of sophistication and customization required, but typically ranges from tens of thousands to hundreds of thousands of euros per unit.

Companies that supply sensor modules aimed at optimizing the gripping force of a robotic arm are, for instance, Baumer, Futek, Honeywell, Omron, Sensata and Schneider. It should be noted that they typically use bulky, complex and costly solutions based on optical or magnetic techniques. The market for robotic sensors is, thus, a high-growth, high-value market.

Mainstream prosthetics manufacturers (e.g., Ottobock, Össur, Open Bionics) still do not largely employ tactile sensing on their prosthetic hands. So far, sensors that measure distributed pressure, vibration, and temperature at the fingertips still see limited adoption in mass-market prosthetic hands because of cost, restricted durability, and clinical/regulatory integration issues (SynTouch Inc., 2025). However, key tailwinds driving the adoption of tactile sensors are: (a) lower sensor costs (SynTouch Inc., 2025), and (b) clinical evidence for improved functional benefits (van der Wijk et al., 2020). Long-term projections (Fortune Business Insights, 2025) see future prosthetics adopting a full closed-loop tactile system (distributed fingertip sensing combined with naturalistic haptic feedback), improving dexterity and embodiment.

While current tactile sensors are still largely interfaced with off-the-shelf electronics, custom readout chips are expected to be more in the forefront. This is due to both emerging highly multiplexed readout demands (Wettels et al., 2008) and constrained energy requirements for edge operations. The chips and electronic modules supplied to that market may be low-volume products, compared to consumer electronics chips, but they nevertheless are an economically viable market due to the high value of the end products.

1.6 Book Overview

This book focuses on pioneering research work in the area of fully-integrated chip-based electronic skin. The book covers the end-to-end design of an electronic skin system. From the intricate design of the tactile sensors, which enable the fine-grained complex force sensing, to the area- and power-efficient per-taxel event-based readout, and to the local and embedded feature extraction/processing for tactile stimuli detection. The goal is to ***leverage the enormous progress in VLSI technology over the last few decades to achieve fully-integrated e-skin system solutions with human-level capabilities.*** This is a departure from a field that has traditionally been sensor fabrication oriented, mainly relegating readouts to off-the-shelf electronics, and relegating the processing to complex offline algorithms.

The book is organized as follows:

- In Chap. 2, a literature review of existing e-skin solutions will be performed. The review covers state-of-the-art works with a system-level end-to-end view, ranging from the tactile sensing transducers, off-the-shelf and recent chip-based readouts, to the processing and data transmission. A recent slew of commercial tactile sensors is covered in this chapter as well. At the end, we will present our case for next-generation fully-integrated e-skin solutions, addressing the limitations of the existing systems.
- In Chap. 3, we will discuss end-to-end design strategies and methodologies to implement next-generation fully-integrated e-skin systems. First, we expound on our pragmatic technology-aware neuromorphic approach, highlighting our approach to not blindly copy biological concepts. Next, we discuss our bioinspired approach to encode tactile signals. We compare our bioinspired event-based approach to conventional frame-based methods. We then show that

our choice to use a spike-based approach lends well with using event-based processing such as SNNs. Such choice of bioinspired processing has been shown to ease the readout demands, reflecting biological observations. We then show novel functions that are important in next-generation fully-integrated e-skin solutions: the integration of sensors capable of high-spatial-resolution texture sensing and slip detection. The engineering challenges and solutions towards per-taxel design, in terms of the chip area and power consumption, are also discussed. Finally, we introduce our multi-modal and multi-scale e-skin: the INTUITIVE and DERMIS approaches.
- In Chap. 4, we will discuss the INTUITIVE chip, highlighting its advantage over prior frame-based e-skin sensor data acquisitions. The INTUITIVE chip, the first to integrate high-density piezoelectric sensors on a commercially-manufactured custom CMOS chip (Fig. 1.6c), highlights its per-taxel spiking readout approach. In this chip, we implement for the first time our pragmatic technology-aware neuromorphic approach, combined with a system-level end-to-end optimization strategy. The detailed circuit implementation and experimental results will then shown.
- In Chap. 5, we present a large-area design to expand beyond the INTUITIVE solution. Using flexible thin-film technology (TFT) (see Fig. 1.6d), we implement our DERMIS chip to achieve per-taxel spiking readout while covering larger sensing areas. With a view to implement grasp-state-aware tactile sensing, we expand on our end-to-end design approach. The detailed circuit implementation and experimental results of the DERMIS chip will then be presented.
- In Chap. 6, our conclusions will be drawn and future outlook will be discussed.

References

Ali, D., Michele, M., Aishwarya, N., Elisabetta, C., Maurizio, V., & Chiara, B. (2021). Artificial bio-inspired tactile receptive fields for edge orientation classification. In *2021 IEEE international symposium on circuits and systems (ISCAS)* (pp. 1–5). IEEE.

Bartlett, H. (2023). *Synthetic touch: Haptic data is revolutionizing consumer products*. USC Viterbi — News [Online]. Available: https://viterbischool.usc.edu/news/2023/01/synthetic-touch-haptic-data-is-revolutionizing-consumer-products/ [Accessed: 24 Sep 2025].

Bartolozzi, C., Natale, L., Nori, F., & Metta, G. (2016). Robots with a sense of touch. *Nature Materials, 15*, 921–925.

Bergner, F., Dean, E., & Cheng, G. (2020). Design and realization of an efficient large-area event-driven e-skin. *Sensors, 20*, 1965.

BMW. (2024). *A touch of the future with bmw*. Online article [Online]. Available: https://www.bmw.com/en/freude/a-touch-of-the-future-with-bmw.html [Accessed: 24 Sep 2025].

Boutry, C., Negre, M., Jorda, M., Vardoulis, O., Chortos, A., Khatib, O., & Bao, Z. (2018). A hierarchically patterned, bioinspired e-skin able to detect the direction of applied pressure for robotics. *Science Robotics, 3*, eaau6914.

Canatu. (2018) *Canatu and continental cooperate for 3d touch in automotive interiors*. Online press release [Online]. Available: https://canatu.com/company/carbon-age-news/canatu-and-continental-cooperate-for-3d-touch-in-automotive-interiors/ [Accessed: 24 Sep 2025].

Canepa, G., Petrigliano, R., Campanella, M., & De Rossi, D. (1998). Detection of incipient object slippage by skin-like sensing and neural network processing. *IEEE Transactions on Systems, Man, and Cybernetics, Part B (Cybernetics), 28*(3), 348–356.

Chen, G., Cao, H., Conradt, J., Tang, H., Rohrbein, F., & Knoll, A. (2020). Event-based neuromorphic vision for autonomous driving: A paradigm shift for bio-inspired visual sensing and perception. *IEEE Signal Processing Magazine, 37*(4), 34–49.

Ciancio, A., Cordella, F., Barone, R., Romeo, R., Bellingegni, A., Sacchetti, R., Davalli, A., Di Pino, G., Ranieri, F., Di Lazzaro, V., Guglielmelli, E., & Zollo, L. (2016). Control of prosthetic hands via the peripheral nervous system. *Frontiers in Neuroscience, 10.*

Clark, N. (2025). *Tesla demonstrates optimus gen 2 dexterity, recalls 2m vehicles.* The Robot Report [Online]. Available: https://www.therobotreport.com/tesla-demonstrates-optimus-gen-2-dexterity-recalls-2m-vehicles/ [Accessed: 24 Sep 2025].

Corniani, G., & Saal, H. (2020). Tactile innervation densities across the whole body. *Journal of Neurophysiology, 124.*

Curcio, C., & Allen, K. (1990). Topography of ganglion cells in human retina. *The Journal of Comparative Neurology, 300*, 5–25.

Dahiya, R. (2019). E-skin: From humanoids to humans [point of view]. *Proceedings of the IEEE 107*(2), 247–252.

Dahiya, R., Yogeswaran, N., Liu, F., Manjakkal, L., Burdet, E., Hayward, V., & Jörntell, H. (2019). Large-area soft e-skin: The challenges beyond sensor designs. *Proceedings of the IEEE, 107*(10), 2016–2033.

Dawood, A. B., Godaba, H., Ataka, A., & Althoefer, K. (2020). Silicone-based capacitive e-skin for exteroception and proprioception. In *2020 IEEE/RSJ international conference on intelligent robots and systems (IROS)* (pp. 8951–8956).

Delhaye, B. P., Jarocka, E., Barrea, A., Thonnard, J.-L., Edin, B., & Lefèvre, P. (2021). High-resolution imaging of skin deformation shows that afferents from human fingertips signal slip onset. *eLife, 10*, e64679.

Du Bois, D., & Du Bois, E. F. (1916) Clinical calorimetry: Tenth paper—a formula to estimate the approximate surface area if height and weight be known. *Archives of Internal Medicine, 17, 6*(2), 863–871.

Fortune Business Insights. (2025). *Tactile sensor market size, share & growth forecast 2025–2032.* https://www.fortunebusinessinsights.com/tactile-sensor-market-111378, Accessed 10 Nov 2025.

GelSight. (2020). *Digitizing the finger: Startup gelsight hones measurement system that could speed up aerospace work.* Forbes [Online]. Available: https://www.gelsight.com/forbes-digitizing-the-finger-startup-gelsight-hones-measurement-system-that-could-speed-up-aerospace-work/ [Accessed: 24 Sep 2025].

Grand View Research (2024). *Tactile sensors — electronic skin market, 2024–2030.* https://www.grandviewresearch.com/horizon/statistics/electronic-skin-market/sensors/tactile-sensors/global, Market segment page.

Gujar, S., & Das, S. (2025) *Tactile sensors market size—by type, product type, by technology, by application analysis, share, growth forecast, 2025–2034.* Industry Report GMI10509, Global Market Insights, Inc. [Online]. Available: https://www.gminsights.com/industry-analysis/tactile-sensors-market [Accessed: 24 Sep 2025].

Hay, E., & Pruszynski, A. (2020). Orientation processing by synaptic integration across first-order tactile neurons. *PLOS Computational Biology, 16*, e1008303.

Holgado, A. C., Piga, N., Tomo, T. P., Vezzani, G., Schmitz, A., Natale, L., & Sugano, S. (2019). Magnetic 3-axis soft and sensitive fingertip sensors integration for the icub humanoid robot. In *IEEE-RAS international conference on humanoid robots (Humanoids).*

Indiveri, G., Linares-Barranco, B., Hamilton, T. J., van Schaik, A., Etienne-Cummings, R., Delbruck, T., Liu, S.-C., Dudek, P., HÄfliger, P., Renaud, S., Schemmel, J., Cauwenberghs, G., Arthur, J., Hynna, K., Folowosele, F., SAÏGHI, S., Serrano-Gotarredona, T., Wijekoon, J., Wang, Y., & Boahen, K. (2011). Neuromorphic silicon neuron circuits. *Frontiers in Neuroscience, 5-2011.*

Janotte, E., Bamford, S., Richter, O., Valle, M., & Bartolozzi, C. (2022) Neuromorphic capacitive tactile sensors inspired by slowly adaptive mechanoreceptors. In *2022 20th IEEE interregional NEWCAS conference (NEWCAS)* (pp. 119–123).

Jarocka, E., Pruszynski, J. A., & Johansson, R. S. (2021). Human touch receptors are sensitive to spatial details on the scale of single fingerprint ridges. *Journal of Neuroscience, 41*(16), 3622–3634.

Jarumethitanont, W., et al. (2024). Development of low-cost pressure mapping device to evaluate force distribution for seat cushion modification. *Scientific Reports, 14*(1), Article 21804.

Johansson, R., & Flanagan, J. (2009). Coding and use of tactile signals from the fingertips in object manipulation tasks. *Nature Reviews. Neuroscience, 10*, 345–359.

Johansson, R., & Vallbo, A. (1979). Tactile sensibility in the human hand: Relative and absolute density of four types of mechanoreceptive units in glabrous skin. *The Journal of Physiology, 286*, 283–300.

Johansson, R., & Westling, G. (1984). Roles of glabrous skin receptors and sensorimotor memory in automatic control of precision grip when lifting rougher or more slippery objects. *Experimental brain research. Experimentelle Hirnforschung. Expérimentation cérébrale, 56*, 550–64.

Jörntell, H., Bengtsson, F., Geborek, P., Spanne, A., Terekhov, A., & Hayward, V. (2014). Segregation of tactile input features in neurons of the cuneate nucleus. *Neuron, 83*(6), 1444–1452.

Lee, W. W., Tan, Y. J., Yao, H., Li, S., See, H., Hon, M., Ng, K., Xiong, B., Ho, J., &Tee, B. (2019a). A neuro-inspired artificial peripheral nervous system for scalable electronic skins. *Science Robotics, 4*, eaax2198.

Lee, Y., Park, J., Choe, A., Cho, S., Kim, J., & Ko, H. (2019b). Mimicking human and biological skins for multifunctional skin electronics. *Advanced Functional Materials, 30*, 1904523.

Lesniak, D. R., Marshall, K. L., Wellnitz, S. A., Jenkins, B. A., Baba, Y., Rasband, M. N., Gerling, G. J., & Lumpkin, E. A. (2014). Computation identifies structural features that govern neuronal firing properties in slowly adapting touch receptors. *eLife, 3*, e01488.

Lichtsteiner, P., Posch, C., & Delbruck, T. (2008). A 128× 128 120 db 15 μs latency asynchronous temporal contrast vision sensor. *IEEE Journal of Solid-State Circuits, 43*(2), 566–576.

Mackevicius, E., Best, M., Saal, H., & Bensmaia, S. (2012). Millisecond precision spike timing shapes tactile perception. *The Journal of Neuroscience: The Official Journal of the Society for Neuroscience, 32*, 15309–15317.

Mao, Q., Liao, Z., Yuan, J., & Zhu, R. (2024). Multimodal tactile sensing fused with vision for dexterous robotic housekeeping. *Nature Communications, 15*, Article 6871.

Mead, C. (1989). *Analog VLSI and neural systems*. Addison-Wesley.

Mordor Intelligence. (2025). Tactile sensor market size & share analysis — growth trends and forecast (2025–2030). Industry Report, Mordor Intelligence [Online]. Available: https://www.mordorintelligence.com/industry-reports/tactile-sensor-market [Accessed: 24 Sep 2025].

Munde, S. (2025). Tactile sensor market research report by application (robotics, consumer electronics, healthcare, automotive, industrial automation), by sensor type (resistive sensors, capacitive sensors, piezoelectric sensors, optical sensors, magnetic sensors), by technology (analog sensors, digital sensors, smart sensors, microelectromechanical systems), by end use (aerospace, manufacturing, healthcare, automotive, consumer electronics) and by regional (north america, europe, south america, asia pacific, middle east and africa)—forecast to 2034. Industry Report MRFR/SEM/32122-HCR, Market Research Future [Online]. Available: https://www.marketresearchfuture.com/reports/tactile-sensor-market-33961 [Accessed 24 Sep 2025].

Murray, S. J., Medinger, J., Schmitz, J., Balkır, S., & Hoffman, M. W. (2020). Toward a low power e-skin interface system on a chip for taxel arrays. In *2020 IEEE international symposium on circuits and systems (ISCAS)* (pp. 1–5).

Navaraj, W., & Dahiya, R. (2019). Fingerprint-enhanced capacitive-piezoelectric flexible sensing skin to discriminate static and dynamic tactile stimuli. *Advanced Intelligent Systems, 1*(7), 1900051.

Osborn, L., et al. (2014). Tactile feedback in upper limb prosthetic devices using flexible textile force sensors. *Bioengineering conference (paper archived on PMC)*.

Othman, W., Lai, Z.-H. A., Abril, C., Barajas-Gamboa, J. S., Corcelles, R., Kroh, M., & Qasaimeh, M. A. (2022). Tactile sensing for minimally invasive surgery: Conventional methods and potential emerging tactile technologies. *Frontiers in Robotics and AI, 8*.

Pednekar, S. (2025). In-depth industry outlook: Tactile sensor market size & forecast. Industry Report ID 183957, Verified Market Research [Online]. Available: https://www.verifiedmarketresearch.com/product/tactile-sensor-market/ [Accessed: 24 Sep 2025].

Platkiewicz, J., Lipson, H., & Hayward, V. (2016). Haptic edge detection through shear. *Scientific Reports, 6*, 23551.

Pruszynski, A., & Johansson, R. (2014). Edge-orientation processing in first-order tactile neurons. *Nature Neuroscience, 17*.

Rasouli, M., Chen, Y., Basu, A., Kukreja, S., & Thakor, N. (2018). An extreme learning machine-based neuromorphic tactile sensing system for texture recognition. *IEEE Transactions on Biomedical Circuits and Systems*, 1–13.

Romeo, R. A., & Zollo, L. (2020). Methods and sensors for slip detection in robotics: A survey. *IEEE Access, 8*, 73027–73050.

Rongala, U. B., Mazzoni, A., Spanne, A., Jörntell, H., & Oddo, C. M. (2020). Cuneate spiking neural network learning to classify naturalistic texture stimuli under varying sensing conditions. *Neural Networks, 123*, 231–246.

Rosa, M. B. A., & Kraft, M. (2024). Microengineered flexible pressure sensors with sacrificial molding layer: A novel fabrication approach for improved performance. *Proceedings, 97*, 1.

Saal, H., & Bensmaia, S. (2014). Touch is a team effort: Interplay of submodalities in cutaneous sensibility. *Trends in Neurosciences, 37*.

Saal, H., Wang, X., & Bensmaia, S. (2016). Importance of spike timing in touch: an analogy with hearing? *Current Opinion in Neurobiology, 40*, 142–149.

Safa, A., Van Assche, J., Alea, M. D., Catthoor, F., & Gielen, G. G. (2022). Neuromorphic near-sensor computing: From event-based sensing to edge learning. *IEEE Micro, 42*(6), 88–95.

Schmitz, J. A., Sherman, J. M., Hansen, S., Murray, S. J., Balkır, S., & Hoffman, M. W. (2019). A low-power, single-chip electronic skin interface for prosthetic applications. *IEEE Transactions on Biomedical Circuits and Systems, 13*(6), 1186–1200.

Skedung, L. (2012). *Tactile perception: Role of friction and texture*. Ph.D Thesis, KTH Royal Institute of Technology, Stockholm, Sweden, 2012.

Sullivan, J. C., Mitchinson, B., Pearson, M. J., Evans, M., Lepora, N. F., Fox, C. W., Melhuish, C., & Prescott, T. J. (2012). Tactile discrimination using active whisker sensors. *IEEE Sensors Journal, 12*(2), 350–362.

Sundaram, S., Kellnhofer, P., Li, Y., Zhu, J.-Y., Torralba, A., & Matusik, W. (2019). Learning the signatures of the human grasp using a scalable tactile glove. *Nature, 569*, 7758.

SynTouch Inc. (2025). Robotics and prosthetics — tactile sensor technology. https://syntouchinc.com/robotics/, Accessed 10 Nov 2025.

TacterionGmbH. (2023). Tacterion, munich-based sensor technology company, raises series a funding from two global industrial players to facilitate collaboration on future-generation products. Online press release [Online]. Available: https://www.pressebox.com/pressrelease/tacterion-gmbh/tacterion-munich-based-sensor-technology-company-raises-series-a-funding-from-two-global-industrial-players-to-facilitate-collaboration-on-future-generation-products/boxid/1062390 [Accessed: 24 Sep 2025].

Tasker, J. F., Frazer, J., Ferranti, G., & Matthews, J. C. F. (2024). A bi-cmos electronic photonic integrated circuit quantum light detector. *Science Advances, 10*(20), eadk6890.

Van Assche, J., Carlino, M. F., Alea, M. D., Massaioli, S., & Gielen, G. (2023). From sensor to inference: end-to-end chip design for wearable and implantable biomedical applications. In *2023 IEEE biomedical circuits and systems conference (BioCAS)* (pp. 1–5).

van der Wijk, U., et al. (2020). Sensory feedback in hand prostheses: A prospective study. *Frontiers in Neuroscience, 14*, 663. Accessed 10 Nov 2025.

Virone, M., Lopes, P., Rocha, R. P., de Almeida, A. T., & Tavakoli, M. (2021). Dynamic hand gesture recognition using a stretchable multi-layer capacitive array, proximity sensing, and a svm classifier. In *2021 IEEE/RSJ international conference on intelligent robots and systems (IROS)* (pp. 7183–7188).

Wettels, N., Santos, V., Johansson, R., & Loeb, G. (2008). Biomimetic tactile sensor array. *Advanced Robotics, 22*, 829–849.

Willemet, L., Kanzari, K., Monnoyer, J., Birznieks, I., & Wiertlewski, M. (2021). Initial contact shapes the perception of friction. *Proceedings of the National Academy of Sciences, 118*(49), e2109109118.

XELA Robotics. (2025). uskin tactile sensor—use cases. Online. Industrial applications incl. Hitachi and DENSO WAVE [Online]. Available: https://www.xelarobotics.com/use-cases [Accessed: 13 Sep 2025].

Yang, Y.-J., Cheng, M.-Y., Chang, W.-Y., Tsao, L.-C., Yang, S.-A., Shih, W.-P., Chang, F.-Y., Chang, S.-H., & Fan, K.-C. (2008). An integrated flexible temperature and tactile sensing array using pi-copper films. *Sensors and Actuators A: Physical, 143*(1), 143–153. Micromechanics Section of Sensors and Actuators (SAMM), based on contributions revised from the Technical Digest of the IEEE 20th International Conference on Micro Electro Mechanical Systems (MEMS 2007).

Yoon, J., Wang, Z., Kim, K., Wu, H., Ravichandran, V., Xia, Q., Hwang, C., & Yang, J. (2018). An artificial nociceptor based on a diffusive memristor. *Nature Communications, 9*(1). Publisher Copyright: © The Author(s) 2018.

Zhou, F., & Chai, Y. (2020). Near-sensor and in-sensor computing. *Nature Electronics, 3*, 664–671.

Chapter 2
Literature Review of E-skin Solutions

Abstract In this chapter, a literature review of existing e-skin solutions will be performed. The review covers state-of-the-art works with a system-level ***end-to-end*** view, ranging from the tactile sensing transducers, the readout electronics, to the processing, and the data transmission strategies. This end-to-end literature review reflects the focus of our work on the end-to-end design of e-skin systems, with the goal of implementing a complete tactile sensor system to approximate the incredible tactile perception capabilities of the human skin. Furthermore, it looks at aspects of prior e-skin works via the lens of its potential towards implementing the fundamental principles of human skins, as discussed in Sect. 1.1: (a) shear and normal force sensing, (b) event-based signal encoding, (c) high-density sensor-to-afferent connections, and (d) spatiotemporal data compression.

Keywords Electronic skin · Tactile sensor readout · Neuromorphic processing · End-to-end design

2.1 Transducers

The research into electronic skins so far has mainly focused on developing the tactile sensors able to transduce normal force (Dahiya et al., 2014; Navaraj & Dahiya, 2019). The development of tactile transducers (referred to as such in this chapter to avoid confusion with *sensors*, which can also refer to the system as a whole) is not trivial for many reasons. One reason is that in contrast to existing biological-inspired sensors such as artificial retinas (Lichtsteiner et al., 2008) and cochleae (Yang et al., 2016), artificial skin needs to interact with the environment that it is sensing, thus making it prone to wear and tear. Secondly, the current fabrication processes that allow sensors to have skin-like properties, such as force transduction, flexibility, and viscoelasticity, limit the spatial resolution achievable by the sensors.

Prior literature reviews covering e-skin transducers often categorize existing works according to the type of transducer used. E-skin transducers can be neatly categorized into: (a) capacitive (Boutry et al., 2018; Cheng et al., 2010), (b) piezoelectric (Dahiya et al., 2014; Gupta et al., 2018), (c) piezoresistive (Lee et al.,

M. D. Alea, G. Gielen, *Chips That Feel*, Analog Circuits and Signal Processing,
https://doi.org/10.1007/978-3-032-26584-5_2

2017), (d) magnetic (Bhirangi et al., 2021), and (e) vision-based (Yuan et al., 2015) types. In this literature review, we will discuss existing e-skin transducers via the following: (a) the capability to detect multi-axial forces, (b) the potential for CMOS integration, and (c) the potential for scaling in terms of spatial resolution and the number of sensors. In this way, we can highlight several aspects of prior works which are relevant towards our goal of implementing high-density power-efficient tactile sensor systems, similar to human skins.

2.1.1 Shear Force Transducers

A key aspect of the human skin sensing mechanisms often overlooked in literature is the multi-axial (e.g., shear and normal) force sensing component performed in the human fingertips. Early tactile transducer implementations (Christ & Sanderson, 1982) often simplify the mechanisms of human skin sensing into a two-dimensional normal force transducer array. Human skin, however, due to its mechanical compliance, is able to detect complex tensile, compressive, and torsional forces during object handling. Therefore, more recent e-skin transducer work tried to incorporate this observation into their designs. Capacitive shear sensors are common implementations of shear-capable transducers (Boutry et al., 2018; Cheng et al., 2010; Yousuf et al., 2022). This is due to the fact that lateral forces applied on a capacitor plate would result in a change in the overlap area between the bottom and the top plates. This is especially true for capacitive sensors with viscous dielectric materials, such as polydimethylsiloxane (PDMS). Similar to how minute changes in the shear forces change the capacitive sensor area overlap, minute shear-force-induced changes in the angle and alignment of reflectors result in a change in the intensity of the reflected light. Recent light-based shear force sensors (Khamis et al., 2018; Ulloa et al., 2022) exploit this principle in a miniaturized form factor by embedding photodiodes, microlenses, and light incidence detectors inside small pillars. However, despite the miniaturization, shear-capable tactile force transducers still have larger pitch than their simpler normal-force-only counterparts. In summary, the requirement for complex force detection typically complicates the transducer to be used, reducing the sensing spatial resolution and increasing the number of mechanically-moving parts. We will utilize capacitive shear force sensing in Chap. 5.

2.1.2 Potential for CMOS Integration

As discussed in Sect. 1.3, transducer-to-readout electronics integration presents several benefits related to: (a) bandwidth/noise improvements, (b) improvements in the mechanical robustness related to wiring *wear and tear*, and (c) enabling *true* dense per-taxel conversions. MEMS-based transducers, not only lend themselves well for extreme miniaturization, but also have the potential for CMOS chip integration. MEMS-based transducers, owing to their precise fabrication process (Wang & Lu, 2024; Kane et al., 2000), achieve: (a) a large number of integrated taxels on a chip (e.g., 32 × 32 Wang & Lu, 2024 and 64 × 64 Kane et al., 2000), (b) high sensitivity (e.g., 0.67 mV/kPA (Wang & Lu, 2024) and 1.59 mV/kPA (Kane et al., 2000)), and small pitches (175 μm Wang & Lu, 2024 and 300 μm Kane et al., 2000). The floating structure of MEMS-based transducers however results in major limitations such as a reduced force dynamic range, which could result in permanent damage when exposed to large transient forces (Spinner et al., 2006). Prior tactile piezoelectric transducers (Pinna et al., 2017) have been shown to be fully-compatible with CMOS processes, resulting in arrays with a small pitch and the opportunity for direct interfacing with electronics integrated on the same substrate. Piezoelectric oxide semiconductor field effect transistors (POSFETs) (Dahiya et al., 2014) present a step further towards CMOS integration by directly embedding a piezoelectric material (commonly, polyvinylidene fluoride—trifluoroethylene barium titanate or PVDF-TrFE) inside MOSFET structures. Such an approach eliminates the use of charge amplifiers (Pinna et al., 2017) for sensor signal conditioning, while still providing the necessary high input impedance and charge-to-voltage conversion. Flexible POSFETs (Gupta et al., 2018) pave the way for high-spatial-resolution sensing integrated on bendable substrates, crucial in applications wherein a certain radius of curvature needs to be attained. A common limitation with tactile piezoelectric transducers, however, is the high voltage required (thickness-dependent; $\sim$100 V/μm) to perform a post-fabrication procedure called *poling*. Poling aligns the internal dipoles inside the PVDF-TrFE layer using high voltages, which risks damaging the fabricated transistors, as well. Furthermore, prior fabricated POSFETs have only been demonstrated on non-commercial fabrication processes, resulting in POSFETs and corresponding electronics with a coarse feature size ($L = 12$ μm Dahiya et al., 2014; Gupta et al., 2018). The integration of piezoelectric sensors on a high-density commercial CMOS chip is demonstrated in our first prototype (see Chap. 4).

2.1.3 Scaling Potential

As touched upon in Sect. 1.2.1, human fingertips are populated by a large number of mechanoreceptors with sub-millimeter spatial resolution, resulting in fine-grained sensing of features such as textures or edge orientations. Prior e-skin works therefore attempt to approximate this property by scaling both the number of sensors and the spatial resolution. A common strategy to achieve density is by using active crossbar arrays of sensing units (Lee et al., 2017; Rosa & Kraft, 2024). Crossbar arrays are implemented by inserting a tactile sensing material (e.g., dielectric) directly in-between the row and column lines, forming the sensors' top and bottom electrodes. Such an approach also allows frame-based row and column addressing to acquire the sensor signals in a sequential manner. Piezoresistive crossbar arrays in Lee et al.

(2017) allow the detection of transient macro-features such as the indentation of an object. The modeled large number of transducers (4096), combined with the high spatial resolution (100 μm) is used to generate a tactile image which depicts the indentation of a curved object. This transducer array, however, was not fabricated, but was only emulated using finite element modeling. An active array of flexible microstructured capacitive pressure sensors was explored in Rosa and Kraft (2024). Fabricated with photolitographic processes, small conical vertical microstructures were embedded in a 10×10 array of $120 - \mu m$ pitch, improving the sensitivity. In Wu et al. (2013), vertical zinc oxide (ZnO) nanowires (NW) sandwiched between indium tin oxide electrodes exhibit piezotronic properties (the alignment of non-mobile ions in the crystal) when exposed to axial strains as a result of mechanical deformation. A 92×92 array with around $100 - \mu m$ pitch allows the detection of macrofeatures such as an object indentation or letter patterns. Transducer arrays, however, still need readout electronics to get valuable data out of it. As will be alluded to in the following sections, scaling the active array size with frame-based readout acquisitions could result in system-level power and bandwidth penalties.

2.1.4 Summary: Array Upscaling of Complex Force Sensors Demands Sensor Miniaturization and Integration

Replicating the rich tactile perception of human fingertips during object manipulation requires equally power-efficient, high-spatial-resolution sensing and localized processing. Consequently, the number and density of transducers must approach those of human mechanoreceptors. Achieving such fine-grained spatial resolution further necessitates transducers with minimal pitch. Current tactile sensors (Rasouli et al., 2018; Lee et al., 2019), however, fail to reach sub-millimeter spatial resolution.Photolithography-based fabrication offers a promising pathway toward this goal, enabling the precise fabrication and scalable realization of high-resolution sensor arrays. In parallel, sensor-to-readout integration provides a strategy that not only addresses challenges in signal conditioning and mechanical robustness, but also scales favorably by enabling further miniaturization and higher sensor counts. Finally, multi-axial force sensing remains inherently more complex than normal-force detection, making miniaturization particularly challenging. Capacitive shear sensors, owing to their relatively simple structure, present an attractive solution for this miniaturization challenge.

2.2 Readout

Reflecting the wide variety of prior e-skin systems, the readout implementations in literature vary widely, as well. As depicted in Fig. 2.1, a general trend observed

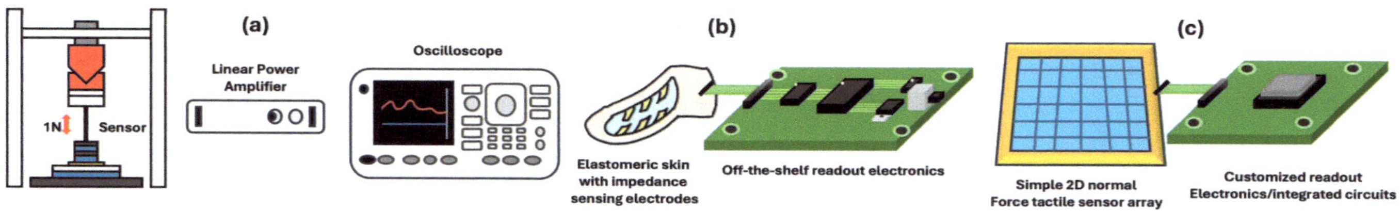

Fig. 2.1 A general trend observed is that more complex tactile sensors tend to have less custom readout electronics, likely related to the observation that a lot of the research in the e-skin field tends to be dominated by sensor-driven approaches. More custom-electronics-oriented readout approaches then tend to have relatively simple sensors. (**a**) Bench instrument readout (Adapted from Wu et al., 2017), (**b**) Off-the-shelf readout electronics (including image of the fluid-based tactile sensor) (Image depicts system from Wettels et al., 2008), (**c**) Custom readout (Image depicts system from Aygun et al., 2019)

is that the more complex tactile transducers tend to have less custom readout electronics. This is likely related to the observation that a lot of the research in the e-skin field tends to be dominated by transducer-driven approaches. The focus is to demonstrate the tactile transducer properties, thus, bench lab instruments are normally used. In contrast, more custom-electronics-oriented readout approaches tend to have simpler transducers. This is possibly borne out of the fact that more complex transducers could complicate readout approaches, while simpler transducers are easily suitable for a proof-of-concept implementation. Meanwhile, neuromorphic designs, implemented in both off-the-shelf and custom chip readouts, are inspired from the sensor-to-spike conversion performed in human skin. Figure 2.4 further on will summarize prior readout work in terms of the readout power consumption versus the raw data throughput/data rate.

2.2.1 *Neuromorphic Readouts*

Neuromorphic tactile sensors mimic the force-to-spikes conversion performed in mechanoreceptor-afferent tandems in human skin. Prior neuromorphic implementations combine both off-the-shelf (Lee et al., 2019) and custom CMOS solutions (Janotte et al., 2022; Caviglia et al., 2017). Biological spike trains have been hypothesized to encode the tactile stimuli in both: (a) their rate codes (Birznieks et al., 2001; Jenmalm et al., 2003) and (b) their precise timing (Johansson & Birznieks, 2004; Richmond & Wiener, 2004; Saal et al., 2016). In rate-based encoding, the sensor value is proportional to the average population (i.e., from multiple channels) firing rate over a certain period, which typically relies on long-term spike count integration. Spike-timing-based encoding, on the other hand, relies on the precise order and timing of spatiotemporal spike trains to encode the signal. Since it relies on the initial bursts of a few spikes, low-latency mechanisms such as first-to-spike encoding have been used (Lee et al., 2017).

One of the simplest ways to convert a continuous-time transducer signal into a spike train (or any discrete-time signal) is to embed the sensor within a voltage-controlled oscillator (VCO). Implementations of this kind typically combine the transducer and the readout on a single substrate, typically flexible (Fig. 2.2a). The force-sensitive resistors (FSR) in Kim et al. (2018) and Tee et al. (2015) modulate the voltage bias of a simple ring oscillator, implemented on a flex substrate. In Tan et al. (2020), the output of a VCO is connected to an LED for transmission to an optical fiber. However, by inserting the resistor directly into the path to the supply voltage, the resulting resistance-to-frequency conversion is very nonlinear. A more linear implementation would have been to convert the piezoresistance to a current, which is then used to bias a current-starved VCO. A more complicated issue would be due to the rail-to-rail voltage dependency with the sensor value, complicating the logic level translation of the subsequent logic stage.

Implementations mimicking more biologically-plausible sensor-to-spike conversions typically borrow concepts from existing analog-based neuromorphic proces-

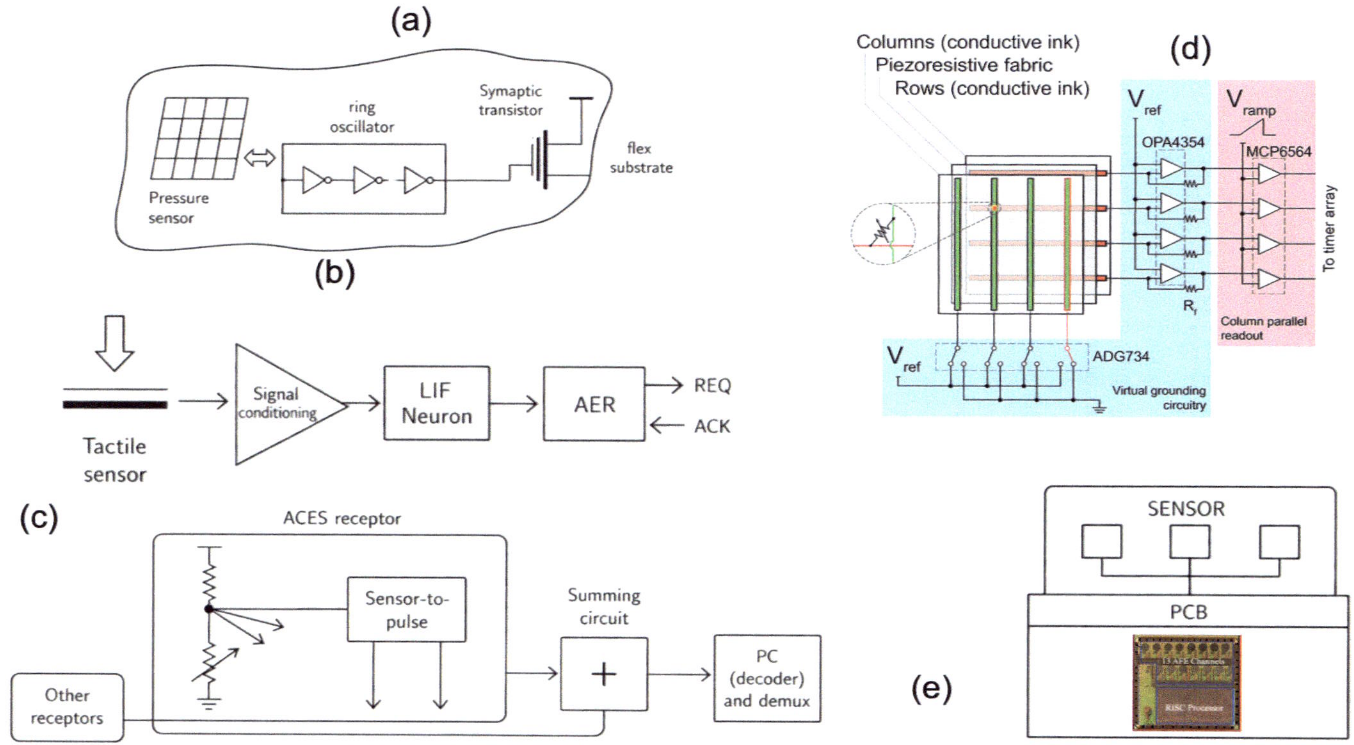

Fig. 2.2 (**a**) Force-sensitive resistors (FSR) modulate the voltage bias of a simple ring oscillator, implemented on flex substrates (Adapted from Kim et al., 2018), (**b**) Neuromorphic readout using LIF neurons on chip (Adapted from Caviglia et al., 2017), (**c**) Neuromorphic readout using off-the-shelf hardware (Adapted from Lee et al., 2019), (**d**) Off-the-shelf readout (Adapted from Lee et al., 2017, licensed under CC BY 4.0), (**e**) Custom ASIC readout (Adapted from Schmitz et al., 2019)

sors, specifically spike-to-current low-pass filters (Chicca et al., 2014). In Janotte et al. (2022), Caviglia et al. (2017), and Simmich et al. (2023), CMOS implementations of leaky integrate-and-fire (LIF) neuronal models replace synaptic inputs from other neurons by the output of a POSFET (Caviglia et al., 2017; Simmich et al., 2023) or a capacitive sensor (Janotte et al., 2022). The transducer output, when converted into a current, is then integrated into the membrane (implemented with a capacitor). This results in spike-train outputs that resemble biological neurons more closely, compared to VCO-based implementations (Fig. 2.2b). Alternative implementations (Zhang et al., 2019) skip the spike generation, converting only the sensing output into a post-synaptic current to be integrated by a succeeding stage (i.e., a membrane).

Sensor-to-spike conversion is performed in Lee et al. (2019) to compress multiple transducer outputs to a spike train in a single conductor. This is achieved by quantizing the output of an FSR (configured in a resistive divider) by an on-board ADC of an *Attiny*20 microcontroller (Fig. 2.2c). The outputs of several channels of the same configuration are then coupled to a single conductive line.When implemented with a single conductive fabric, wear resistance is enhanced due to the inherent redundancy of the single-conductor approach (i.e., a large continuous conductive sheet provides multiple parallel conduction paths). It also reduces the necessary wiring (i.e., wiring is already embedded in the fabric) to transmit the sensed input in the array, which further degrades robustness to mechanical contact (see Sect. 1.3). The single-conductor approach, however, has an inferior throughput compared to more standard neuromorphic interfaces such as the Address Event Representation (AER) (Boahen, 2000). Furthermore, the technique requires a complex upstream decoding algorithm at the receiving end to decouple the spikes from multiple receptors.

Neuromorphic readouts excel in mimicking biologically-plausible firing behavior. Prior off-the-shelf readout approaches use power-inefficient readout, which do not scale well with increasing the number of sensors. Large-area implementations on custom exotic substrates, meanwhile, could be prone to reduced yield and fabrication defects. Chip-based neuromorphic approaches, on the other hand, due to their strict biomimicry approach, are power-inefficient due to both current-mode operation and the additional complexity to faithfully mimic neuronal firing. Implementing current-mode low-pass filters, although they consume small subthreshold currents in each taxel, could easily accumulate when the number of taxels scales up. Furthermore, the low-pass filter output's weak-inversion operation is more susceptible to mismatch.

2.2.2 *Off-the-shelf Electronics*

A step up from typical bench instrument-based readouts, e-skins geared towards real-life demonstrations tend to have commercial off-the-shelf electronics.

Capacitive e-skin sensors tend to have crossbar arrays, interfaced with row-column addressing. In Lee et al. (2017), a 64×64 piezoresistive fabric array is interfaced with column-based slope ADCs (Fig. 2.2d). Columns of piezoresistive sensing elements are connected to resistive inverting amplifiers (OPA4354) and comparators (MCP6564) with a slope input. Timers capture the elapsed time from the start of the slope to when the threshold is reached, resulting in a digitized code of the piezoresistor value.

Piezoelectric sensors in Pinna et al. (2017) use multiple channels of charge amplifiers, bandpass filters, multiplexed to a single ADC resulting in a $0.01 - pC$ charge resolution. The piezoelectric sensor in Navaraj and Dahiya (2019) is coupled and biased to a MOS gate with a resistive divider, resembling a common-source gain stage before quantization by an ADC on-board a $PIC18f45k22$ microcontroller. POSFETs (Dahiya et al., 2014), on the other hand, intrinsically provide the necessary high input impedance and charge-to-voltage conversions. The quantization of the voltage signal, however, is performed only by standard bench-level instrumentation.

Neuromorphic off-the-shelf readouts typically convert transducer information into spikes at the software level. Thus, off-the-shelf ADCs are used for the readout, with overdesigned precision and speed as opposed to the low-precision nature of biological neurons. Frame-based readouts typically use 10- to 12-bit ADCs (Rasouli et al., 2018; Sankar et al., 2022) but can go as high as 24-bits (Rongala et al., 2019). Thus, acquisition datarates can go from 21.6 kHz (Rasouli et al., 2018) to 146 kHz (Rongala et al., 2019) despite the limited number of sensors (between 4 to 9). Consequently, for large sensor arrays such an approach become inefficient.

The implementation of skin sensor readouts has mainly been relegated to using custom off-the-shelf centralized electronics with overdesigned precision and speed (Lee et al., 2017, 2019; Ulloa et al., 2022). While this allows for quick measurements and characterizations of skin sensors in existing designs, the readout circuits often have bulky footprints and a large power consumption. Although this may be sufficient for existing skin sensor designs with less than 100 sensors and coarse spatial resolution, this becomes an issue for emerging large-area electronic skin with fine spatial resolution and two to three orders of magnitude more sensors, such as those seen in human skin. Recent chip-based e-skin readout approaches try to remedy this issue.

2.2.3 *Chip-based Approaches*

Driven by the stringent power consumption requirement (for reduced battery size and/or longer battery life) of e-skin interfaces for prosthetic applications (Schmitz et al., 2019; Murray et al., 2020), single-chip designs have started gaining ground. In Schmitz et al. (2019) and Murray et al. (2020), multiple channels of capacitive transimpedance amplifiers, and 10-bit SAR ADCs are used to convert charge inputs from off-chip tactile piezoelectric transducers (Fig. 2.2e). The presented chip offers a holistic microcontroller-like approach to tactile readouts, integrating timers, memories, serial interface controllers, and a RISC processor on a single chip. This approach allows flexible software-based reconfiguration of the readout operation depending on the target application. Such an approach, however, is fundamentally limited as the sensor count (currently at 13 Schmitz et al., 2019) is scaled up to reach human-level density. Aside from large readout frontend chip areas, which occupies 3.12 mm^2, the sensor count is fundamentally-limited by the number of bondpads and wires required for the transducer-to-readout connections.

Prior chip-based readout approaches address the transducer scaling issue by addressing both: (a) the readout area (Damilano et al., 2017; Han et al., 2024), and (b) the number of required physical transducer-to-readout connections (Chen et al., 2024; Aygun et al., 2019). Multiplexing multiple inputs into a single or a few channels has traditionally been used to reduce the readout area, ammortizing a single or a few ADCs onto multiple inputs each. Multiplexing, however, increases the bandwidth requirements for the succeeding converter, thus setting an upper limit to the number of multiplexed inputs. Thus, multi-channel readout implementations are still constrained by the chip area. On the other hand, as discussed previously, reducing the required number of physical transducer-to-readout connections increases the system's robustness to wear and tear.

Time-based sensor readout architectures (Gielen et al., 2020) offer a promising way in reducing the readout area by leveraging as much digital circuitry as possible. CMOS scaling is largely motivated by putting more digital circuits on a single chip, thus, exploiting digital circuits on the readout chain is beneficial for area reduction. As compared to conventional amplitude-based quantization readouts, which require power-hungry and large amplifiers, and do not scale well with the shrinkage of CMOS feature sizes and the related reduction of supply voltage, time-based frontends offer the possibility for chip area reduction. While *digitizing* (transformed to a rail-to-rail signal) the sensor input early on in the frontend chain, typically via VCOs, the sensor signal is converted into phase information. Then, digital circuits can be exploited to directly quantize this phase information in a time-to-digital approach. The relative simplicity of embedding a capacitive sensor inside a VCO renders time-based architectures attractive for tactile readout applications. In Damilano et al. (2017), a capacitance-to-digital converter (CDC) using a replica-based VCO-based architecture is used to achieve 0.0176 mm^2 area in a 0.13 μm CMOS process. Meanwhile, multiplexed CDCs in Han et al. (2024), using a simplistic first-order architecture of an open-loop VCO followed by a counter, achieve a low readout channel area of 0.015 mm^2 in a 0.18 μm CMOS process.

Methods to reduce the wiring complexity between the off-chip sensors and the chip-based readouts have been proposed in Chen et al. (2024) and Aygun et al. (2019). In Chen et al. (2024), the number of connections required is reduced to 16 sensors per wire via multiplexing. As compared to conventional multiplexing which incurs noise penalties in noise-shaping architectures, a Walsh code division multiplexing (CDM) technique is used for the parallel acquisition. The large sensor per wire metric reported is a result of increasing the number of parallel columns addressed by a single channel. Thus, the number of bondpads and wiring to connect both the row and column lines to the chip still limit the potential for scaling up the number of sensors. In Aygun et al. (2019), the number of required interfaces have been reduced by a factor of M/N through a spatial compression sensing algorithm (Fig. 2.1c). Compressed sensing (Baraniuk, 2007) works on the idea that, if an input signal is sampled with a measurement matrix ϕ that is K-sparse in the spatial or the frequency domain, the original input (with N samples) can be reconstructed with just $M > K \times log_2(N/K)$ samples, where $M << N$. The use of a random and binary measurement matrix ϕ allowed a simple compressed sensing implementation on a large-area thin-film technology (TFT) using early ZnO-TFT NMOS switches. While the number of physical connections is reduced due to the compressed sensing, the readout in Aygun et al. (2019) is still, however, performed by another silicon CMOS chip, precluding a fully-integrated TFT solution.

2.2.4 *Summary 1: Neuromorphic Readouts Open the Door for Neuromorphic Processing, But Current Designs Are Insufficient for Scaled Designs*

Neuromorphic readouts offer the following advantages in the following: (a) the inherent temporal data compression due to the event-based conversion, and (b) the inherent compatibility of such spiking readout with energy-efficient neuromorphic processing. Prior neuromorphic readouts (Lee et al., 2019; Rongala et al., 2017), however, are mostly implemented with custom off-the-shelf solutions, which does not scale well with increasing the number of sensors. Prior chip-based neuromorphic approaches, on the other hand, with their strict biomimicry approach are power-inefficient due to both the current-mode operation they use and the additional complexity to faithfully mimic neuronal firing functionality. Furthermore, its typical weak-inversion operation is inherently prone to device mismatches and variations.

2.2.5 *Summary 2: Scaling Tactile Sensor Array Sizes Demands Aggressive Readout Constraints*

As detailed in Sect. 2.1, increasing the number of sensor units is essential to reach the density of the tactile receptors in human skin. However, scaling the sensor array also necessitates a proportional increase in the number of readout channels. Consequently, readout architectures must have compact footprints and a low power consumption. Chip-based implementations are particularly well-suited to address these requirements, benefiting from the continuous scaling of CMOS technologies. Another challenge arises from the increasing number of physical sensor-to-readout interconnects, which are susceptible to wear and tear degradation under mechanical stress. Approaches such as time-based architectures, code-division multiplexed readouts, and compressed sensing alleviate some of these constraints in terms of increased readout channel counts and reduced sensor-to-readout wiring, though they currently lack the sparse temporal sampling typical for neuromorphic readouts (Fig. 2.3).

2.3 Processing and Data Transmission

Borrowing from its biological counterpart, e-skins have generally been used to sense various force- and temperature related parameters to perform a classification task. In the human somatosensory system, spike trains from the primary afferents travel to the spinal cord for further processing before going straight to the brain. The distributed nature of e-skin presents many challenges with respect to:

- transmitting the high-throughput sensed information to a central processing node over relatively long distances, and
- performing low-latency and low-power classification tasks at the central node.

The advent of machine learning (ML) techniques in computation offers energy reduction over conventional digital processing (Boahen, 2017). Inspired by biological neurons, artificial neural networks (ANNs) are computational models that perform neural functions at their basic sense (Gerstner et al., 2014): (a) sum weighted input, then (b) get the output after passing through a non-linear activation function. The network model is trained using error-minimization algorithms such as backpropagation/gradient descent (Rumelhart et al., 1986). Spiking neural networks (SNNs) use spikes (discrete events) instead of continuous values, thus, are closer to biology. SNNs, while being a relatively newer development, are now slowly becoming at par with ANN classification performance (Zambrano & Bohte, 2016) while requiring less energy. Common training approaches are inspired from prior ANN learning methods, of using surrogate gradient (Safa et al., 2023) to replace the non-differentiable activation function of a spiking neuron. Although, more recent techniques use biologically-plausible learning algorithms such as spike-timing-dependent plasticity (STDP) (Safa et al., 2024). SNNs show to be more power-efficient due to their sparse and asynchronous activity, scaling well with the sparsity of the input signals (Indiveri & Sandamirskaya, 2019).

2.3.1 Conventional Frame-based Approaches

Conventional *frame-based* signal acquisition and processing approaches have been explored in Murray et al. (2020) and Schmitz et al. (2019). Instead of the event-based sensing and processing observed in prior e-skin work, the work in Murray et al. (2020) utilized a 32-bit RISC-V microcontroller unit (MCU) with a hardware

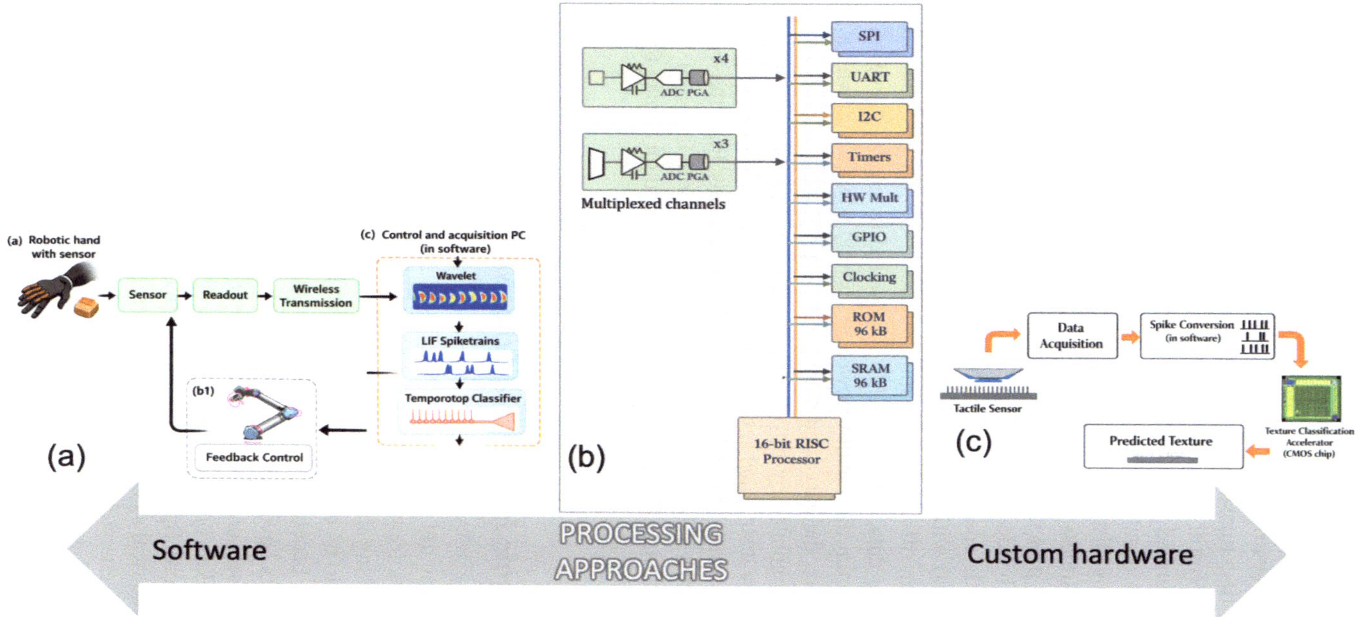

Fig. 2.3 (**a**) Software-based classification (Reproduced from Navaraj & Dahiya, 2019, licensed under CC-BY-4.0), (**b**) Custom chip with integrated RISC-V processor, for flexible software support (Adapted from Schmitz et al., 2019), (**c**) Custom chip using an accelerator approach called extreme learning machine (ELM) (Adapted from Rasouli et al., 2018). A key difference between the approaches in (**b**) & (**c**) is that the RISC-v based approach uses a general-purpose programmable central processing unit (CPU) core for flexible software execution, whereas an accelerator approach implements dedicated hardware for a specific task to achieve better performance at the expense of flexibility

neural processing unit (NPU) to provide support for an artificial neural network (ANN) implementation. While the actual ANN implementation in Murray et al. (2020) has not been discussed extensively, the work described using the network for a gesture classification task. Instead of transmitting high-throughput tactile sensor data, integrating a processor on-chip localizes the processing and eliminates or reduces the need for data transmission. However, the work in Murray et al. (2020) and Schmitz et al. (2019) still implemented the MCU on a separate $65nm$ chip, while the function of the integrated 16-bit RISC-V described in the text was not discussed.

Due to the distributed nature of sensing and the challenge of transmitting raw tactile frames over long distances, it does not come as a surprise that recent e-skin work leverages wireless transmission (Hao et al., 2024). Similar to body area networks of the recent past, the work in Hao et al. (2024) leverages multiple low-power nodes with embedded sensing and transmitters to form a tactile sensing network. The approach enables a flexible placement of sensing nodes (often placed in the fingertips) without the wiring complexity for both data and power connections. Instead, an off-chip coil is multiplexed to choose between receiving power and transmitting data. Since the chip is powered by energy harvesting, the chip has to work with minimal standby power and is only powered up with a wake-up receiver. Furthermore, the harvested nature of its supply constrains its circuits to work with just $0.4V$, regulated by an on-chip low-dropout regulator (LDO). While the work approaches the distributed sensing dilemma with a novel wireless network approach, the chip's constrained energy budget fundamentally limit the system to a single sensor approach (Hao et al., 2024). The bulky form factor of a single node currently prevents high-density tactile sensing, as needed on human fingertips (Jarocka et al., 2021).

2.3.2 Neuromorphic Event-based Processing

Similar to biological neural networks, SNNs use asynchronous spikes as inputs instead of synchronous frames. A sensor-to-spike encoder is, thus, highly compatible to generate spiking input for SNNs, while also being power-efficient in its sensing operation due to its event-driven nature (Safa et al., 2022). Aside from the low-power processing benefits of asynchronous spike-based classifiers in SNNs, signal-to-spike encoders have been leveraged for the data compression of the high-throughput raw tactile sensor data (Van Assche et al., 2023).

In human skin, spikes or action potentials travel all the way to the cuneate neurons in the spinal cord. Spikes do no travel with direct electrical continuity (like in a single wire in electrical systems) but through relays of neurotransmitter release and post-synaptic ion channel activation (Gerstner et al., 2014). This mechanism is emulated in early electronic systems via an AER interface, where distributed individual spikes are transmitted from one chip to another (inspired from the neuron relays) via asynchronous handshaking and address-based timestamps. Preserving the asynchronous nature of the spike generation and transmission, the AER interface communicates via request (REQ) and acknowledge (ACK) lines. The neural spike origin is then preserved by encoding the specific neuron as a distinct address in both X and Y ($ADDRX/ADDRY$). Neuromorphic processors, as well as a prior neuromorphic chip-based e-skin sensor (Caviglia et al., 2017) utilize this principle. The work in Lee et al. (2019) simplified this approach by compressing the spike train into a single conductor, in exchange of a reduced throughput.

Spike-based encoders are inherently event-based. They can be leveraged to sample the tactile stimuli, which can be time-sparse. Such property however is not fully leveraged in prior e-skin work (Rongala et al., 2017; Navaraj & Dahiya, 2019), which performs the spike conversion after the ADC acquisition and in software, in contrast to on-chip signal encoders in Caviglia et al. (2017) and Janotte et al. (2022). Furthermore, SNN-based classification is often performed in software. Thus, sensor data are acquired by an off-the-shelf readout board and stored in memory for later processing (i.e., classification tasks). Much of the e-skin solutions in this category are motivated by utilizing neuromorphic hardware to closely mimic biological sensing and processing principles. For example, the work in Rongala et al. (2020) explores the use of biologically-plausible cuneate neuron behavior to perform the same secondary processing performed in the spinal cord. Thus, the focus was clearly not how to perform both signal-to-spike conversion and spike transmission efficiently in hardware.

Similarly, offline software-based SNN-based approaches to detect slip have been proposed in literature (Follmann et al., 2024). Detecting slip in human skin is not a trivial task. It involves both: (a) complex multi-axial force sensing, and (b) a low-latency slip detection algorithm. This is addressed in Follmann et al. (2024) by using an off-the-shelf multi-axis force-torque sensor ($JR3$) and an SNN model emulated in neuromorphic processors like SpiNNaker (Furber et al., 2014) and Intel's Loihi (Davies et al., 2018). The event-based slip detection in Nakagawa-Silva et al. (2019) generates a spike every time the input signal exceeds a dynamically-changing threshold to achieve a $< 85ms$ detection latency. In Juárez-Lora et al. (2022), a Kalman filter is implemented using an SNN model, overcoming the need for an accurate modeling of the system dynamics and a proper characterization of uncertainties with a learning-based approach. Kalman filters, commonly used in tactile slip detection for robotics (Dutta et al., 2024), combine a predictive motion model with noisy sensor readings to get robust predictions of incipient slip. Software-based spike encoding, however, fundamentally increases the slip detection latency as sensor data still have to be sampled periodically by a fixed-rate ADC. Unless sampled very high at the expense of huge system power consumption, fixed-sampling-rate ADCs are often inadequate for the fast dynamics of an ongoing slippage event.

In Rasouli et al. (2018), an extreme learning machine (ELM) chip has been described to perform spike-based texture data classification on custom hardware. The proposed hardware mimics on-chip the computational properties of neurons by integrating weighted inputs and biases, then using a nonlinear activation to generate the output spikes. The chip accepts input spikes encoded in digital form, although it is unclear what kind of interface was used and how the spikes are represented. The chip forms a rate-coded representation of the input spike train using a counter, then transforming this digital value into an analog value via current DACs. The output currents are then fed onto 128 hidden-layer neurons, exploiting the random mismatch between the current mirrors to perform analog-based computing. The output generation from the hidden layer, however, is performed off-chip. Furthermore, the spike inputs are still generated in software, neglecting the data compression benefits of direct signal-to-spike conversion before ADC acquisition and data transmission.

2.3.3 Summary: Neuromorphic Sensing and Processing—Benefits Beyond Software Implementations

To match the same temporal fidelity of human skin sampling (Mackevicius et al., 2012), conventional frame-based readouts need to have high sampling rates when interfacing a large taxel array. Furthermore, data transmission of high-throughput raw tactile sensor data over relatively large distances can be challenging, as the large number of channels and high sampling rates lead to very high data rates that strain bandwidth, power, and signal integrity over long interconnects. Thus, borrowing inspiration from the human skin, local processing and/or data compression algorithms are beneficial in reducing the throughput. By converting continuous-time sensor signals into discrete-time spikes, neuromorphic sensing offers a solution due to its event-based sampling, which captures the fine temporal dynamics of tactile signals while also adapting to periods of low activity. The sparse sampling inherently provides data compression, which provides system-level benefits, especially due to the inherently distributed nature of tactile sensing. Current neuromorphic readout approaches, however, typically still depend on off-chip software-based spike conversion, implying that signal acquisition continues to be performed by power-inefficient, frame-based analog-to-digital converters. As a result, frames of potentially highly redundant tactile data are acquired and discarded after software-based conversion, leading to significant inefficiencies. Beyond this system-level overhead, software-based spike conversion also introduces much larger latency, making it unsuitable for time-sensitive, real-time applications.

2.4 Commercial Tactile Sensors

E-skin represents an emerging application with significant untapped market potential (see Sect. 1.5). It has driven the development of several state-of-the-art commercial tactile sensors in recent years. Commonly envisioned for object manipulation, commercial tactile sensors tend to focus on slip detection. The high dimensionality of the complex multi-axial forces to be measured to detect slip often leads to the use of either optical-based or fluid-impedance-based sensing in the commercially available shear-capable sensors. In the commercial sensor of (Yuan et al., 2015), a high-resolution camera is used to capture the displacement of dots on an elastic polymer when the latter gets in contact with an object. Such an approach yields a snapshot with high spatial resolution of the contact forces, allowing the use of image entropy extraction to predict slip. While this achieves excellent slip detection, such an approach results in excessively large acquisition data rates (frames of RGB videos) and complicated upstream processing (entropy extraction), which limits its scalability towards larger-area sensing.

The commercial sensor from (Khamis et al., 2021) uses a similar optical-based approach to detect lateral forces. The deflection of tiny pillars mounted on the sensor changes the direction of where the reflected infrared light is beamed onto four photodiodes. The individual light intensities measured by the photodiodes are then used to infer the lateral force applied to the pillars.

While a low-cost open-source alternative is presented in Lambeta et al. (2020), it inherits all the said issues with optical-based shear sensing. The commercial sensor in Wettels et al. (2008) achieves composite force detection by allowing an internal liquid to deform during contact, from which deformity is measured through the impedance between an electrode array. However, this fingertip-based sensor costs around \$10k (not including the regular \$2k to refill the liquid), thus making it cost-prohibitive for large-scale implementation. A magnetic-based approach to detect the 3D deformation during contact is presented in Robotics (n.d.). The magnetic nature of the sensing, however, makes it susceptible to crosstalk between the sensors and interference from nearby magnetic fields.

In summary, in contrast to earlier e-skin works, commercial tactile sensors, put a lot of focus on shear and normal force sensing. This is related to our earlier discussion that complex force sensing performed by human skin is a crucial mechanism required during grasp manipulation and object property identification. The lack of system integration, especially with regards to their use of off-the-shelf readout, results in low power efficiency and does not scale favorably when the number of sensors is increased.

2.5 Summary of Prior E-skin Works

A comprehensive summary of the leading prior work in e-skin designs is shown in Table 2.1. In line with the overall end-to-end strategy in this book, the table summarizes and compares prior e-skin work with an end-to-end view. The table covers sensor details: the type of sensor, the number of taxels, and the nature of the measured force/pressure. The table also covers readout-related implementation details: the readout architecture, the input parameter sensed by the readout, the supply voltage, and the sensing sampling rate. Finally, the table presents processing and data transmission details such as: the data transmission protocol, the processing model complexity (as reflected by the neuron counts), and the pattern/tactile stimuli classified. The leading prior e-skin works summarized in this table show gaps

Table 2.1 Summary of prior e-skin work: end-to-end

	Dahiya et al. (2014)	Caviglia et al. (2017)	Kim et al. (2018)	Navaraj and Dahiya (2019)	Lee et al. (2017)	Sankar et al. (2022)	Rongala et al. (2020)	Rasouli et al. (2018)
Research group	University of Glasgow	University of Genoa	Stanford	University of Glasgow	National University of Singapore	Johns Hopkins	Lund University	Nanyang University of Singapore
Description	POSFET	POSFET (events)	Organic device	Tempotron classifier	Fast classifier	Soft robotic fingertips	Cuneate classifier	ELM
Pattern classified	None	None	Braille	Textures	Contact with ball	13 textures	10 textures (scan)	Grasp planning
System bus to processing/NN	Serial	AER	Async. parallel	Bluetooth	Async. parallel	Serial	Async. parallel	No bus
Readout architecture (before processing/NN)	Multiplexed ADC (COTS)	LIF neuron	VCO	2 ADC (COTS)	1 ADC + pre-process	1 ADC + pre-process	1 ADC + pre-process	1 ADC + pre-process
Taxel circuits	Current mirror, buffer	OTA, LIF	VCO/polymer MOS	Current source, switch	Comparator, timer	N/A	N/A	Switches; resistors
Taxel array output	Voltage	Frequency	Frequency	Voltage	Spike	ADC code	ADC code	ADC code
Taxel area	0.9 mm × 0.6 mm	0.075 mm × 0.227 mm	30 mm × 8 mm[a]	Not specified	2.29 mm × 2.29 mm	2 mm × 2 mm	1 mm wide	1 mm wide
Number of sensors	16 (4×4)	1	6 (2×3)	1 static, 1 dynamic	4096 (64×64)[a]	9 (3×3)	16 (8 used)	6 (2×3)
Sensor array sampling rate	30 kHz	CT	Not specified	200 Hz	5.2 kHz	13 kHz	~24 Hz per array	300 Hz
Sensor type	POSFET	POSFET	Resistive	Capacitive + Piezoelectric	Piezoresistive	Piezoresistive fabric	Piezoresistive MEMS	Piezoresistive
Measured force	Dynamic (0.01–3 N; ~102.4 mV/N)	Dynamic (approx. 0.1–1.5 N)	Static (1–80 kPa; ~1 Hz/kPa[a])	Static (< 100 Pa to 2.5 kPa) + Dynamic	Dynamic (software)	Static	Static + Dynamic (software)	Static + Dynamic (software)
Supply voltage	10 V rail (±5 V)	3.3 V	10.5 V rail (2–8.5 V)	Not specified	Not specified	Not specified	Not specified	2.1 V & 0.6 V
Number of neurons	N/A	1 (LIF)	17 (current soma)	1 (tempotron)	128 software LIF	N/A (SVM)	32 Izhikevich; 25 cuneate (software)	~128 (ELM)

[a]Emulated array

Table 2.2 Somatosensory properties studied in neuroscience and bioinspired in electronic skins

Somatosensory property	Studied in neuroscience papers	Mimicked in e-skin
Rate coding of mechanoreceptors	Birznieks et al. (2001) and Jenmalm et al. (2003)	Kim et al. (2018), Tee et al. (2015), Caviglia et al. (2017), and Tan et al. (2020)
Relative spike timing of mechanoreceptors	Johansson and Birznieks (2004), Richmond and Wiener (2004), and Saal et al. (2016)	Navaraj and Dahiya (2019), Lee et al. (2017), Lee et al. (2013), and Rongala et al. (2017)
Vibration encodes texture/roughness through fingerprint ridges	Weber et al. (2013), Scheibert et al. (2009), Fagiani et al. (2011), Lieber and Bensmaia (2019), Harvey et al. (2013), Vickery et al. (2020), Mackevicius et al. (2012), and Manfredi et al. (2014)	Park et al. (2015) and Oddo et al. (2011)
Visual–tactile fusion for hand control	Johansson and Flanagan (2009), Jonas et al. (2019), and Adams et al. (2013)	Taunyazov et al. (2020), Navaraj et al. (2019), Bologna et al. (2013), and Cui et al. (2016)
2nd-order neuron classification	Jörntell et al. (2014), Rongala et al. (2018), Hayward et al. (2014), Johansson and Flanagan (2009), and Bengtsson et al. (2013)	Rongala et al. (2020), Navaraj and Dahiya (2019), Bologna et al. (2011), and Lee et al. (2013)
1st-order neurons encode stimulus property	Pruszynski and Johansson (2014), Hay and Pruszynski (2020), and Jarocka et al. (2021)	Ali et al. (2021) and Kim et al. (2018)
Shear force sensing informs slip detection	Delhaye et al. (2021), Johansson and Westling (1984), and Khamis et al. (2014)	Khamis et al. (2018) and Khamis et al. (2021)

in terms of the following: (a) limited number of sensors, (b) the use of power-inefficient commercial off-the-shelf readout, and (c) the use of complex upstream tactile classification algorithms. Our research aims to design power-efficient readout that scales well with the increasing number of tactile sensors, while local embedded processing offloads part of the computational load at the edge, simplifying upstream processing.

A second table associating prior bioinspired e-skin work to the fundamental human skin sensing biological property is shown in Table 2.2. It links each tactile sensing or neural mechanism (e.g., rate coding, spike timing, vibration-based texture encoding, visual–tactile fusion, shear-force-based slip detection) to representative e-skin references, highlighting how insights from basic sensory science have translated into artificial tactile sensing.

Figure 2.4 compares our INTUITIVE work with prior tactile sensor readouts, both off-the-shelf and ASIC-based. Increasing the number of sensors to increase the spatial sampling fidelity of the tactile stimuli with the increased sensing sampling

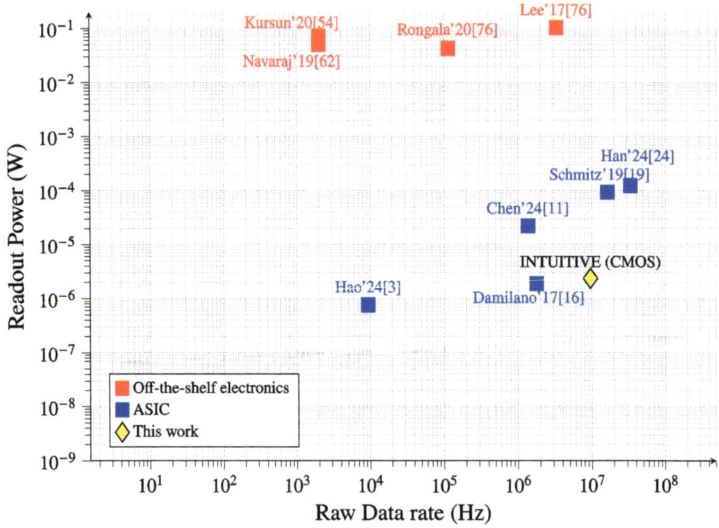

Fig. 2.4 Readout power versus raw data rate across prior tactile sensor readout and this work

rate due to the fine temporal dynamics of tactile stimulation events, generates very large raw throughputs for tactile sensors. A general trend with end-to-end sensing systems (Van Assche et al., 2023) is that larger raw throughputs require a larger power consumption, owing to the increased sampling and data transmission rate. Hence, temporal compression of raw data would result in a more power-efficient design despite a large raw throughput. Figure 2.5 compares our slip-sensing DERMIS system with prior slip-capable tactile sensors. Prior tactile sensors capable of slip detection (Romeo et al., 2017; Fernández et al., 2014; Cheng et al., 2010; Su et al., 2015; Leslie et al., 2023) have been implemented almost exclusively using commercial off-the-shelf components, reflecting the relatively recent development of this capability. Several orders of magnitude improvement in the power-efficiency is a result of both a fully-integrated approach and optimized end-to-end design.

In summary, e-skin design requires a holistic end-to-end approach from the transducer design, to the readout implementation, and to the data processing and transmission. Transducer-focused designs, as we have shown, neglect the complexity of the readout interface design, typically relegating them to commercial off-the-shelf electronics. Thus, real-life deployment of such systems is difficult. On the other hand, readout-electronics-focused designs, particularly recent chip-based implementations, use simple, coarse spatial resolution transducers for their proof-of-concept. Such e-skin systems overlook the complex, fine-grained tactile information captured by biological skin, which is also crucial for neuroprosthetic and robotic applications. Based on biological observations, Table 2.3 summarizes the hardware requirements for texture and slip sensing. Our research fills the gap in between these prior approaches.

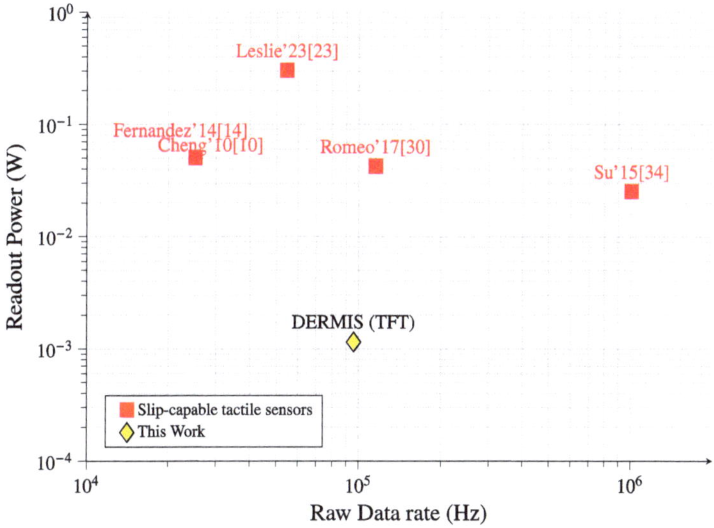

Fig. 2.5 Readout power versus raw data rate across prior slip-capable tactile sensor readout and this work

Table 2.3 Hardware requirements for texture and slip sensing based on their biological equivalents

Requirement	Texture sensing (Hay & Pruszynski, 2020)	Slip sensing (Delhaye et al., 2021)
Pitch	400 µm	500 µm
Power[a]	2.4 nW	2.4 nW
Sampling rate	Adaptive event-based (max BW: 500 Hz)	Adaptive event-based (max BW: 400 Hz)
Local processing	Complex receptive fields	Grasp-state-adaptive force extraction
Sensing area	Localized small area	Large area
Force sensed	Dynamic force (vibration)	Normal and shear forces
Biological equivalent	Merkel cells/Pacinian corpuscles	Meissner corpuscles

[a] Power values derived in Appendix A

What is end-to-end design? End-to-end design means you consider the full chain: how the real-world is sensed, how the sensed parameter is digitized, how it is processed, how it is communicated, and how ultimately the inference or decision is made. Rather than optimizing each stage in isolation, the design aims to co-optimize across these layers to achieve better overall system performance and efficiency.

End-to-end design in e-skin

E-skin design encompasses several disciplines, namely, sensor design, readout development, and algorithm optimizations. The complexity and breadth of e-skin design has understandably relegated much of the prior research in this field to address small subsets of the whole e-skin challenge. A significant chunk of the field has traditionally been sensor fabrication oriented, mainly relegating readouts to off-the-shelf electronics, and relegating the processing to complex offline algorithms. An emerging chip-based approach to tackle the e-skin problem has so far worked with relatively simpler sensors, disregarding many of the important sensing aspects in e-skin like shear force sensing. Algorithm-oriented e-skin research, on the other hand, improves tactile stimuli classification through various approaches, commonly using neuromorphic sensing and processing. The algorithm focus, however, has settled with power-inefficient off-the-shelf readout that do not scale well in real-life large-array applications.

While this book does not claim to comprehensively address all facets of e-skin design, it aims to take a modest first step towards a more integrated treatment of sensing, readout, and processing.

Prior designs: where can we improve on? Several examples illustrate how the absence of an end-to-end perspective can limit e-skin performance:

- The e-skin research field is ripe with various implementations of flexible tactile sensors (Gupta et al., 2018). Such an approach approximates the skin's compliance to force input, allowing for bending and stretching of the sensor itself. However, a key obstacle towards large-scale adoption of such approach is the required physical wiring (Sundaram et al., 2019) to connect the sensor to the electronics would introduce the following side-effects: (a) sensor signal degradation and noise coupling, which may be a function of the unintended force or movement applied to the physical connection itself (i.e., movement-induced noise), and (b) the reduced compliance of the sensor due to the sensitive connection itself.
- In Kim et al. (2018), the force-sensitive resistors (FSR) modulate the voltage bias of a simple ring oscillator. This results in the rail-to-rail output of the VCO having a voltage dependency with the sensor value, complicating the logic level translation of the subsequent logic stage.
- Neuromorphic design approaches attempt to closely mimic how biological neurons convert receptor information to spikes. This is commonly done with Izhikevich neurons, but is still mostly done at software level (Rongala et al., 2019). Thus, the readout misses on opportunities for sparse-sampling, which would have power benefits in both conversion and data transmission.

(continued)

- In vision-based tactile sensors (Yuan et al., 2015), the complex force profile when an elastic polymer contacts an object modulates the position of dots, then is captured by a high-resolution camera. Such an approach yields a high spatial resolution snapshot of the contact forces, allowing the use of image entropy extraction to predict slip. While this achieves excellent slip detection, such an approach results in excessively large acquisition data rates and complicated upstream entropy extraction processing.

2.6 The Case for Next-generation Fully-integrated E-skin Solutions

Our literature review yields us the following key insights:

- array size scaling of complex force sensors demands sensor miniaturization and integration,
- scaling tactile sensor array sizes demands aggressive readout constraints,
- neuromorphic readouts open the door for neuromorphic processing, but current designs are insufficient for scaled designs, and
- neuromorphic sensing and processing offer system-level processing and data compression benefits, but prior work are limited to software implementations

To approximate the rich tactile information captured by the human fingertips during object manipulation, it is necessary to achieve the same power-efficient and high-spatial-resolution sensing and local processing. Therefore, transducer counts need to scale to the level of the number and density of mechanoreceptors in the human fingertips. Aside from the sensor count, these transducers also need to have small sensor pitch to approximate the fine-grained sensing with high spatial resolution. However, interfacing increasingly huge tactile transducer counts present challenges for the readout electronics. Furthermore, the complicated wiring needed to interface the array, even with multiplexed implementations, present robustness issues related to the wear and tear brought about by mechanical movements in typical applications. Given these issues, the integration of dense tactile transducers on the same substrate as the electronics is a promising alternative solution. Prior demonstrations of this (Dahiya et al., 2014; Kim et al., 2018), still rely on tactile transducer integration on non-commercial CMOS-capable processes with coarse feature sizes. Implementing dense and power-efficient readout electronics is a challenge in such substrates, increasing the system cost and presenting thermal dissipation issues. Conventional frame-based readouts, to match the same temporal fidelity of human skin sampling (Mackevicius et al., 2012), need to have high sampling rates to interface a large array. Furthermore, data transmission of high-throughput raw tactile sensor data over relatively broad distances could be challenging. Thus, borrowing inspirations from the human skins, local processing and/or data compression algorithms could be beneficial in reducing the throughput.

In summary, e-skin systems benefit from the high-level of integration afforded by VLSI technologies. Per-taxel readout of integrated sensors (i.e., 1:1 sensor-to-readout ratio) presents a pathway towards solving the aforementioned limitations. The sensor-to-readout chip integration supports dense per-taxel connections, enabling *true* per-taxel sparse-sampling conversion (i.e., spikes) as performed in human afferents, while also eliminating the mechanically-fragile sensor-to-readout wiring. To achieve this, integrating a dense tactile sensor array on a custom CMOS readout chip, allows for dense and power-efficient readout with local processing capabilities. However, moving from off-the-shelf solutions to full integration *per se* does not guarantee a one-size-fits-all solution, but rather requires a suite of complementary approaches framed within a system-level, end-to-end perspective. For example, can the miniaturized sensor reduce the required per-taxel area (e.g., smaller input capacitance needs smaller readout capacitances)?

Aside from the inherent power- and area-efficiency improvements due to high levels of integration, custom fully-integrated chip designs open up a plethora of system-level end-to-end optimizations that may be rooted on biological inspirations. However, porting all *techniques* from biology to artificial substrates, does not necessarily guarantee the same optimized solutions. Therefore, our bioinspired approach should follow a pragmatic technology-aware methodology. For example, can the use of biologically-inspired and learning-based SNN classification tolerate the readout circuit non-idealities, thus offering power- and area-optimization opportunities?

Looking forward, we present our case for an end-to-end, fully integrated *e*-skin as a pathway toward next-generation tactile sensing systems. By combining the system-level advantages enabled by high integration with the design flexibility of custom integrated solutions, we move closer to replicating the complex, high-density, and power-efficient sensing capabilities of human skin. Drawing inspiration from biology through an engineering lens, we adopt a pragmatic, technology-aware bioinspired approach. The next chapter elaborates on this methodology in greater detail.

References

Adams, M. J., Johnson, S. A., Lefèvre, P., Levesque, V., Hayward, V., André, T., & Thonnard, J.-L. (2013). Finger pad friction and its role in grip and touch. *Journal of the Royal Society Interface, 10*(80), 20120467.

Ali, D., Michele, M., Aishwarya, N., Elisabetta, C., Maurizio, V., & Chiara, B. (2021). Artificial bio-inspired tactile receptive fields for edge orientation classification. In *2021 IEEE international symposium on circuits and systems (ISCAS)* (pp. 1–5).

Aygun, L. E., Kumar, P., Zheng, Z., Chen, T.-S., Wagner, S., Sturm, J. C., & Verma, N. (2019). Hybrid lae-cmos force-sensing system employing tft-based compressed sensing for scalability of tactile sensing skins. *IEEE Transactions on Biomedical Circuits and Systems 13*(6), 1264–1276.

Baraniuk, R. G. (2007). Compressive sensing [lecture notes]. *IEEE Signal Processing Magazine, 24*(4), 118–121.

Bengtsson, F., Brasselet, R., Johansson, R. S., Arleo, A., & Jörntell, H. (2013). Integration of sensory quanta in cuneate nucleus neurons in vivo. *PLOS ONE, 8*(2), e56630.

Bhirangi, R., Hellebrekers, T., Majidi, C., & Gupta, A. (2021). Reskin: Versatile, replaceable, lasting tactile skins. In *Proceedings of the 5th conference on robot learning (CoRL).*

Birznieks, I., Jenmalm, P., Goodwin, A. W., & Johansson, R. S. (2001). Encoding of direction of fingertip forces by human tactile afferents. *Journal of Neuroscience, 21*(20), 8222–8237.

Boahen, K. (2000). Point-to-point connectivity between neuromorphic chips using address events. *IEEE Transactions on Circuits and Systems II: Analog and Digital Signal Processing, 47*(5), 416–434.

Boahen, K. (2017). A neuromorph's prospectus. *Computing in Science & Engineering 19*, 14–28.

Bologna, L. L., Pinoteau, J., Brasselet, R., Maggiali, M., & Arleo, A. (2011). Encoding/decoding of first and second order tactile afferents in a neurorobotic application. *Journal of Physiology-Paris, 105*(1–3), 25–35.

Bologna, L. L., Pinoteau, J., Passot, J.-B., Gandolfo, J. A., Vogel, J., Vidal, E. R., & Arleo, A. (2013). A closed-loop neurobotic system for fine touch sensing. *Journal of Neural Engineering, 10*(4), 046019.

Boutry, C., Negre, M., Jorda, M., Vardoulis, O., Chortos, A., Khatib, O., & Bao, Z. (2018). A hierarchically patterned, bioinspired e-skin able to detect the direction of applied pressure for robotics. *Science Robotics, 3*, eaau6914.

Caviglia, S., Pinna, L., Valle, M., & Bartolozzi, C. (2017). Spike-based readout of posfet tactile sensors. *IEEE Transactions on Circuits and Systems I: Regular Papers, 64*(6), 1421–1431.

Chen, Y., Cai, T., Kuang, Y., Dong, J., Cheng, Z., Zhao, B., & Luo, Y. (2024). A fully dynamic event-driven capacitive sensor interface circuits based on self-reconfigurable sar capacitance-to-digital conversion for high-density robotic tactile sensing. *IEEE Journal of Solid-State Circuits, 59*(8), 2618–2629.

Cheng, M.-Y., Lin, C.-L., & Yang, Y.-J. (2010). Tactile and shear stress sensing array using capacitive mechanisms with floating electrodes. In *2010 IEEE 23rd international conference on micro electro mechanical systems (MEMS)* (pp. 228–231).

Chicca, E., Stefanini, F., Bartolozzi, C., & Indiveri, G. (2014). Neuromorphic electronic circuits for building autonomous cognitive systems. *Proceedings of the IEEE, 102*(9), 1367–1388.

Christ, J. P., & Sanderson, A. C. (1982). A prototype tactile sensor array. Technical Report, CMU-RI-TR-82–14, Pittsburgh, PA, September 1982.

Cui, Y., Lan, Z., Hu, Y., Zhang, Y., & Huang, T. (2016). *Deep neural networks for tactile understanding*.Preprint. arXiv:1604.04656

Dahiya, R. S., Adami, A., Pinna, L., Collini, C., Valle, M., & Lorenzelli, L. (2014). Tactile sensing chips with posfet array and integrated interface electronics. *IEEE Sensors Journal, 14*(10), 3448–3457.

Damilano, A., Motto Ros, P., Sanginario, A., Chiolerio, A., Bocchini, S., Roppolo, I., Pirri, C. F., Carrara, S., Demarchi, D., & Crepaldi, M. (2017). A robust capacitive digital read-out circuit for a scalable tactile skin. *IEEE Sensors Journal, 17*(9), 2682–2695.

Davies, M., Srinivasa, N., Lin, T.-H., Chinya, G., Cao, Y., Choday, S. H., Dimou, G., Joshi, P., Imam, N., Jain, S., Liao, Y., Lin, C.-K., Lines, A., Liu, R., Mathaikutty, D., McCoy, S., Paul, A., Tse, J., Venkataramanan, G., Weng, Y.-H., Wild, A., Yang, Y., Wang, H., & Chan, B. (2018). Loihi: A neuromorphic manycore processor with on-chip learning. In *Proceedings of the 2018 IEEE international symposium on computer architecture (ISCA)* (pp. 1–14).

Delhaye, B. P., Jarocka, E., Barrea, A., Thonnard, J.-L., Edin, B., & Lefèvre, P. (2021). High-resolution imaging of skin deformation shows that afferents from human fingertips signal slip onset. *eLife, 10*, e64679.

Dutta, A., Burdet, E., & Kaboli, M. (2024). Visuo-tactile based predictive cross modal perception for object exploration in robotics. In *2024 IEEE international symposium on robotic and sensors environments (ROSE)* (pp. 1–7).

Fagiani, R., Massi, F., Chatelet, E., Berthier, Y., & Akay, A. (2011). Tactile perception by friction induced vibrations. *Tribology International, 44*(10), 1100–1110.
Fernández, R., Payo, I., Vazquez, A., & Becedas, J. (2014). Micro-vibration-based slip detection in tactile force sensors. *Sensors, 14*(1), 709–730.
Follmann, J., Gentile, C., Cordella, F., Zollo, L., & Rodrigues, C. (2024). Touch and slippage detection in robotic hands with spiking neural networks. *Engineering Applications of Artificial Intelligence, 136*, 108953.
Furber, S. B., Galluppi, F., Temple, S., & Plana, L. A. (2014). The spinnaker project. *Proceedings of the IEEE, 102*(5), 652–665.
Gerstner, W., Kistler, W. M., Naud, R., & Paninski, L. (2014). *Neuronal dynamics: From single neurons to networks and models of cognition*. Cambridge University Press.
Gielen, G. G., Hernandez, L., & Rombouts, P. (2020). Time-encoding analog-to-digital converters: Bridging the analog gap to advanced digital cmos-part 1: Basic principles. *IEEE Solid-State Circuits Magazine, 12*(2), 47–55.
Gupta, S., Yogeswaran, N., Giacomozzi, F., Lorenzelli, L., & Dahiya, R. (2018). Flexible aln coupled mosfet device for touch sensing. In *2018 IEEE SENSORS* (pp. 1–4).
Han, S., Seong, H., Oh, S., Koo, J., Jin, H., Kim, H. J., Ha, S., & Je, M. (2024). A 72-channel resistive-sensor interface ic with high energy efficiency and a wide input range. In *2024 IEEE international symposium on circuits and systems (ISCAS)* (pp. 1–5).
Hao, H., Richardson, A. G., Ding, Y., Du, L., Allen, M. G., Spiegel, J. V. d., & Aflatouni, F. (2024). 17.10 a 0.4v, 750nw, individually accessible wireless capacitive sensor interface ic for a tactile sensing network. In *2024 IEEE international solid-state circuits conference (ISSCC)* (Vol. 67, pp. 332–334).
Harvey, M. A., Saal, H. P., Dammann, J. F. I., & Bensmaia, S. J. (2013). Multiplexing stimulus information through rate and temporal codes in primate somatosensory cortex. *PLOS Biology, 11*(5), e1001558.
Hay, E., & Pruszynski, A. (2020). Orientation processing by synaptic integration across first-order tactile neurons. *PLOS Computational Biology, 16*, e1008303.
Hayward, V., Terekhov, A. V., Wong, S.-C., Geborek, P., Bengtsson, F., & Jörntell, H. (2014). Spatio-temporal skin strain distributions evoke low variability spike responses in cuneate neurons. *Journal of the Royal Society Interface, 11*(93), 20131015.
Indiveri, G., & Sandamirskaya, Y. (2019). The importance of space and time for signal processing in neuromorphic agents: The challenge of developing low-power, autonomous agents that interact with the environment. *IEEE Signal Processing Magazine, 36*, 16–28.
Janotte, E., Bamford, S., Richter, O., Valle, M., & Bartolozzi, C. (2022). Neuromorphic capacitive tactile sensors inspired by slowly adaptive mechanoreceptors. In *2022 20th IEEE interregional NEWCAS conference (NEWCAS)* (pp. 119–123).
Jarocka, E., Pruszynski, J. A., & Johansson, R. S. (2021). Human touch receptors are sensitive to spatial details on the scale of single fingerprint ridges. *Journal of Neuroscience, 41*(16), 3622–3634.
Jenmalm, P., Birznieks, I., Goodwin, A. W., & Johansson, R. S. (2003). Influence of object shape on responses of human tactile afferents under conditions characteristic of manipulation. *European Journal of Neuroscience, 18*(1), 164–176.
Johansson, R., & Birznieks, I. (2004). First spikes in ensembles of human tactile afferents code complex spatial fingertip events. *Nature Neuroscience, 7*, 170–177.
Johansson, R., & Flanagan, J. (2009). Coding and use of tactile signals from the fingertips in object manipulation tasks. *Nature Reviews. Neuroscience, 10*, 345–359.
Johansson, R., & Westling, G. (1984). Roles of glabrous skin receptors and sensorimotor memory in automatic control of precision grip when lifting rougher or more slippery objects. *Experimental brain research. Experimentelle Hirnforschung. Expérimentation cérébrale 56* (02 1984), 550–564.
Jonas, M. D. E., Spanne, A., Mazzoni, A., Bengtsson, F., Oddo, C. M., & Jörntell, H. (2019). Ubiquitous neocortical decoding of tactile input patterns. *Frontiers in Cellular Neuroscience 13*.

Juárez-Lora, A., García-Sebastián, L. M., Ponce-Ponce, V. H., Rubio-Espino, E., Molina-Lozano, H., & Sossa, H. (2022). Implementation of kalman filtering with spiking neural networks. *Sensors, 22*(22), 8845.

Jörntell, H., Bengtsson, F., Geborek, P., Spanne, A., Terekhov, A., & Hayward, V. (2014). Segregation of tactile input features in neurons of the cuneate nucleus. *Neuron, 83.*

Kane, B., Cutkosky, M., & Kovacs, G. (2000). A traction stress sensor array for use in high-resolution robotic tactile imaging. *Journal of Microelectromechanical Systems, 9*(4), 425–434.

Khamis, H., Redmond, S., Macefield, V., & Birznieks, I. (2014). Tactile afferents encode grip safety before slip for different frictions. In *Proceedings of the 36th annual international conference of the IEEE engineering in medicine and biology society (EMBC 2014): Discovering, innovating, and engineering future biomedicine, 26-30 August 2014, Chicago, Illinois* (pp. 4123–4126). IEEE. IEEE Engineering in Medicine and Biology Society. Annual International Conference; Conference date: 11-07-2022.

Khamis, H., Izquierdo Albero, R., Salerno, M., Shah Idil, A., Loizou, A., & Redmond, S. J. (2018). Papillarray: An incipient slip sensor for dexterous robotic or prosthetic manipulation—design and prototype validation. *Sensors and Actuators A: Physical, 270*, 195–204.

Khamis, H., Xia, B., & Redmond, S. J. (2021). Real-time friction estimation for grip force control. In *2021 IEEE international conference on robotics and automation (ICRA)* (pp. 1608–1614).

Kim, Y., Chortos, A., Xu, W., Liu, Y., Oh, J. Y., Son, D., Kang, J., Foudeh, A. M., Zhu, C., Lee, Y., Niu, S., Liu, J., Pfattner, R., Bao, Z., & Lee, T.-W. (2018). A bioinspired flexible organic artificial afferent nerve. *Science, 360*(6392), 998–1003.

Lambeta, M., Chou, P.-W., Tian, S., Yang, B., Maloon, B., Most, V. R., Stroud, D., Santos, R., Byagowi, A., Kammerer, G., Jayaraman, D., & Calandra, R. (2020). Digit: A novel design for a low-cost compact high-resolution tactile sensor with application to in-hand manipulation. *IEEE Robotics and Automation Letters, 5*(3), 3838–3845.

Lee, W. W., Cabibihan, J., & Thakor, N. V. (2013). Bio-mimetic strategies for tactile sensing. In *SENSORS, 2013 IEEE* (pp. 1–4).

Lee, W. W., Kukreja, S. L., & Thakor, N. V. (2017). Discrimination of dynamic tactile contact by temporally precise event sensing in spiking neuromorphic networks. *Frontiers in Neuroscience, 11-2017.*

Lee, W. W., Tan, Y. J., Yao, H., Li, S., See, H., Hon, M., Ng, K., Xiong, B., Ho, J., & Tee, B. (2019). A neuro-inspired artificial peripheral nervous system for scalable electronic skins. *Science Robotics, 4*, eaax2198.

Leslie, O., Córdova Bulens, D., & Redmond, S. J. (2023). Design, fabrication, and characterization of a novel optical six-axis distributed force and displacement tactile sensor for dexterous robotic manipulation. *Sensors, 23*, 24.

Lichtsteiner, P., Posch, C., & Delbruck, T. (2008). A 128× 128 120 db 15 μs latency asynchronous temporal contrast vision sensor. *IEEE Journal of Solid-State Circuits 43*(2), 566–576.

Lieber, J. D., & Bensmaia, S. J. (2019). High-dimensional representation of texture in somatosensory cortex of primates. *Proceedings of the National Academy of Sciences, 116*(8), 3268–3277.

Mackevicius, E., Best, M., Saal, H., & Bensmaia, S. (2012). Millisecond precision spike timing shapes tactile perception. *The Journal of Neuroscience: The Official Journal of the Society for Neuroscience, 32*, 15309–15317.

Manfredi, L. R., Saal, H. P., Brown, K. J., Zielinski, M. C., Dammann, J. F. I., Polashock, V. S., & Bensmaia, S. J. (2014). Natural scenes in tactile texture. *Journal of Neurophysiology, 111*(9), 1792–1802.

Murray, S. J., Medinger, J., Schmitz, J., Balkır, S., & Hoffman, M. W. (2020). Toward a low power e-skin interface system on a chip for taxel arrays. In *2020 IEEE international symposium on circuits and systems (ISCAS)* (pp. 1–5).

Nakagawa-Silva, F., Alspach, A., Luo, R., Valenzuela, A., & Howe, R. D. (2019). A bio-inspired slip detection and reflex-like suppression method for robotic manipulators. *IEEE Robotics and Automation Letters, 4*(4), 3641–3648.

Navaraj, W. T., & Dahiya, R. (2019). Fingerprint-enhanced capacitive-piezoelectric flexible sensing skin to discriminate static and dynamic tactile stimuli. *Advanced Intelligent Systems, 1*(7), 1900051.

Navaraj, W. R. T., Yogeswaran, N., Liu, Y., Chen, S., Doshi, N., Zou, D., & Dahiya, R. (2019). Fingerprint-enhanced biomimetic tactile-sensing electronic skin, inspired by the structure and functions of the human skin. *Advanced Intelligent Systems, 1*(7), 1900051.

Oddo, C. M., Controzzi, M., Beccai, L., Cipriani, C., & Carrozza, M. C. (2011). Roughness encoding for discrimination of surfaces in artificial active-touch. *IEEE Transactions on Robotics, 27*(3), 522–533.

Park, J., Kim, M., Lee, Y., Lee, H. S., & Ko, H. (2015). Fingertip skin-inspired microstructured ferroelectric skins discriminate static/dynamic pressure and temperature stimuli. *Science Advances, 1*(9), e1500661.

Pinna, L., Ibrahim, A., & Valle, M. (2017). Interface electronics for tactile sensors based on piezoelectric polymers. *IEEE Sensors Journal, 17*(18), 5937–5947.

Pruszynski, A., & Johansson, R. (2014). Edge-orientation processing in first-order tactile neurons. *Nature Neuroscience, 17*.

Rasouli, M., Chen, Y., Basu, A., Kukreja, S., & Thakor, N. V. (2018). An extreme learning machine-based neuromorphic tactile sensing system for texture recognition. *IEEE Transactions on Biomedical Circuits and Systems, 12*(2), 313–325.

Richmond, B. J., & Wiener, M. C. (2004). Recruitment order: a powerful neural ensemble code. *Nature Neuroscience, 7*(2), 97–98.

Xela Robotics. (n.d.). uSkin sensors. Retrieved 9 Nov 2025, from https://xelarobotics.com/en/home

Romeo, R. A., Oddo, C. M., Carrozza, M. C., Guglielmelli, E., & Zollo, L. (2017). Slippage detection with piezoresistive tactile sensors. *Sensors, 17*(8).

Rongala, U. B., Mazzoni, A., & Oddo, C. M. (2017). Neuromorphic artificial touch for categorization of naturalistic textures. *IEEE Transactions on Neural Networks and Learning Systems, 28*(4), 819–829.

Rongala, U. B., Spanne, A., Mazzoni, A., Bengtsson, F., Oddo, C. M., & Jörntell, H. (2018). Intracellular dynamics in cuneate nucleus neurons support self-stabilizing learning of generalizable tactile representations. *Frontiers in Cellular Neuroscience, 12*, 210.

Rongala, U., Mazzoni, A., Spanne, A., Jörntell, H., & Oddo, C. (2019). Cuneate spiking neural network learning to classify naturalistic texture stimuli under varying sensing conditions. *Neural Networks, 123*.

Rongala, U. B., Mazzoni, A., Spanne, A., Jörntell, H., & Oddo, C. M. (2020). Cuneate spiking neural network learning to classify naturalistic texture stimuli under varying sensing conditions. *Neural Networks, 123*, 231–246.

Rosa, M. B. A., & Kraft, M. (2024). Microengineered flexible pressure sensors with sacrificial molding layer: A novel fabrication approach for improved performance. *Proceedings, 97*(1).

Rumelhart, D. E., Hinton, G. E., & Williams, R. J. (1986). Learning representations by backpropagating errors. *Nature, 323*(6088), 533–536.

Saal, H., Wang, X., & Bensmaia, S. (2016). Importance of spike timing in touch: an analogy with hearing? *Current Opinion in Neurobiology, 40*, 142–149.

Safa, A., Van Assche, J., Alea, M. D., Catthoor, F., & Gielen, G. G. (2022). Neuromorphic near-sensor computing: From event-based sensing to edge learning. *IEEE Micro, 42*(6), 88–95.

Safa, A., Corradi, F., Keuninckx, L., Ocket, I., Bourdoux, A., Catthoor, F., & Gielen, G. G. E. (2023). Improving the accuracy of spiking neural networks for radar gesture recognition through preprocessing. *IEEE Transactions on Neural Networks and Learning Systems, 34*(6), 2869–2881.

Safa, A., Ocket, I., Bourdoux, A., Sahli, H., Catthoor, F., & Gielen, G. G. E. (2024). Stdp-driven development of attention-based people detection in spiking neural networks. *IEEE Transactions on Cognitive and Developmental Systems, 16*(1), 380–387.

Sankar, S., Slepyan, A., Iskarous, M. M., Cheng, W.-Y., DeBrabander, R., Zhang, J., Gupta, A., & Thakor, N. V. (2022). Flexible multilayer tactile sensor on a soft robotic fingertip. In *2022 IEEE sensors* (pp. 1–4). IEEE.

Scheibert, J., Leurent, S., Prevost, A., & Debrégeas, G. (2009). The role of fingerprints in the coding of tactile information probed with a biomimetic sensor. *Science, 323*(5920), 1503–1506.

Schmitz, J. A., Sherman, J. M., Hansen, S., Murray, S. J., Balkır, S., and Hoffman, M. W. (2019). A low-power, single-chip electronic skin interface for prosthetic applications. *IEEE Transactions on Biomedical Circuits and Systems, 13*(6), 1186–1200.

Simmich, S., Yilmaz, M., Ashkrizzadeh, R., Petraru, A., & Rieger, R. (2023). A piezoelectric biologically plausible spiking neuron. In *2023 IEEE international symposium on circuits and systems (ISCAS)* (pp. 1–5).

Spinner, S., Bartholomeyczik, J., Becker, B., Doelle, M., Paul, O., Polian, I., Roth, R., Seitz, K., & Ruther, P. (2006). Electromechanical reliability testing of three-axial silicon force sensors. In *Proceedings of the design, test, integration and packaging of MEMS/MOEMS (DTIP 2006)* (pp. 77–82). TIMA Editions.

Su, Z., Hausman, K., Chebotar, Y., Molchanov, A., Loeb, G. E., Sukhatme, G. S., & Schaal, S. (2015). Force estimation and slip detection/classification for grip control using a biomimetic tactile sensor. In *2015 IEEE-RAS 15th international conference on humanoid robots (humanoids)* (pp. 297–303).

Sundaram, S., Kellnhofer, P., Li, Y., Zhu, J.-Y., Torralba, A., & Matusik, W. (2019). Learning the signatures of the human grasp using a scalable tactile glove. *Nature, 569*, 7758.

Tan, H., Tao, Q., Pande, I., Majumdar, S., Liu, F., Zhou, Y., Persson, P., Rosén, J., & Dijken, S. (2020). Tactile sensory coding and learning with bio-inspired optoelectronic spiking afferent nerves. *Nature Communications, 11*.

Taunyazov, T., Sng, W., See, H., Lim, B., Kuan, J., Ansari, A. F., Benjamin C. K. T., & Soh, H. (2020). Event-driven visual-tactile sensing and learning for robots. In *Robotics: Science and systems (RSS), 2020*.

Tee, B. C.-K., Chortos, A., Berndt, A., Nguyen, A. K., Tom, A., McGuire, A., Lin, Z. C., Tien, K., Bae, W.-G., Wang, H., Mei, P., Chou, H.-H., Cui, B., Deisseroth, K., Ng, T. N., & Bao, Z. (2015). A skin-inspired organic digital mechanoreceptor. *Science, 350*(6258), 313–316.

Ulloa, P. M., Bulens, D. C., & Redmond, S. J. (2022). Incipient slip detection for rectilinear movements using the PapillArray tactile sensor. In *2022 IEEE sensors* (pp. 1–4). IEEE.

Van Assche, J., Carlino, M. F., Alea, M. D., Massaioli, S., & Gielen, G. (2023). From sensor to inference: end-to-end chip design for wearable and implantable biomedical applications. In *2023 IEEE biomedical circuits and systems conference (BioCAS)* (pp. 1–5).

Vickery, R. M., Gaitatzis, C., Rigas, A., Ennis, C. M., & Mahns, D. A. (2020). Tapping into the language of touch: using non-invasive stimulation to specify tactile afferent firing patterns. *Frontiers in Neuroscience 14*, 500.

Wang, Y.-C., & Lu, M. S.-C. (2024). High-resolution cmos tactile-proximity sensor array: Design, fabrication, and performance evaluation. *IEEE Sensors Journal, 24*(16), 25468–25475.

Weber, A. I., Saal, H. P., Lieber, J. D., Cheng, J.-W., Manfredi, L. R., III, J. F. D., & Bensmaia, S. J. (2013). Spatial and temporal codes mediate the tactile perception of natural textures. *Proceedings of the National Academy of Sciences of the United States of America 110*(42), 17107–17112.

Wettels, N., Santos, V., Johansson, R., & Loeb, G. (2008). Biomimetic tactile sensor array. *Advanced Robotics, 22*, 829–849.

Wu, W., Wen, X., & Wang, Z. (2013). Taxel-addressable matrix of vertical-nanowire piezotronic transistors for active/adaptive tactile imaging. *Science (New York, N.Y.), 340*.

Wu, T., Liu, G., Zhao, L., Li, X., Wang, X., & Yan, Y. (2017). Flexible piezoresistive tactile sensor based on electrospun graphene/tpu nanofibers for human motion monitoring. *Micromachines, 8*(4), 102.

Yang, M., Chien, C.-H., Delbruck, T., & Liu, S.-C. (2016). A 0.5 v 55 μW 64 $\times$ 2 channel binaural silicon cochlea for event-driven stereo-audio sensing. *IEEE Journal of Solid-State Circuits, 51*(11), 2554–2569.

Yousuf, M., Arya, D. S., Garg, M., Varshney, D., & Singh, P. (2022). Asymmetric capacitive pixel array (e-cap) for mapping of epidermal vectored movement. *IEEE Journal on Flexible Electronics, 1*(2), 134–140.

Yuan, W., Li, R., Srinivasan, M. A., & Adelson, E. H. (2015). Measurement of shear and slip with a gelsight tactile sensor. In *2015 IEEE international conference on robotics and automation (ICRA)* (pp. 304–311).
Zambrano, D., & Bohte, S. M. (2016). *Fast and efficient asynchronous neural computation with adapting spiking neural networks*.
Zhang, C., Ye, W., Zhou, K., Hongye, C., Yang, J.-Q., Ding, G., Chen, X., Zhou, Y., Zhou, l., Li, F., & Han, S.-T. (2019). Bioinspired artificial sensory nerve based on nafion memristor. *Advanced Functional Materials, 29*, 1808783.

Chapter 3
Towards Next-generation Fully-integrated E-skin Solutions

Abstract This chapter presents a two-pronged, technology-aware bioinspired approach for next-generation electronic skins that aims to narrow the gap with biological tactile sensing in spatial resolution, power efficiency, embedded processing, feature extraction, and data compression. Rather than directly imitating biology, the approach selectively adopts biological principles that effectively address key e-skin bottlenecks, particularly through spike-based encoding and true spatiotemporally sparse sampling. The chapter shows that temporally sparse, per-taxel spike-based readout can overcome the limitations of conventional frame-based methods, which remain power-inefficient and difficult to scale despite sparse tactile inputs. By combining sensor miniaturization with integrated piezoelectric and capacitive shear sensors, the proposed architecture improves scalability while reducing wiring, bandwidth, and noise-related constraints. Power modeling and comparative analysis demonstrate that event-based spiking readouts provide a clear power-efficiency advantage over synchronous frame-based conversion, especially during inactive periods and at low signal-to-noise ratios. In addition, the proposed neuromorphic level-crossing analog-to-digital converter, inspired by adaptive neuronal thresholding, enables robust and efficient encoding while relaxing circuit design constraints through compatibility with self-correcting spiking neural network classifiers.

Keywords Spike-based encoding · Event-based readout · Sensor integration · Spiking networks

In the previous chapter, we have discussed prior works on electronic skin solutions, or tactile sensor systems. We have highlighted key limitations in the prior designs and have highlighted an overall lack of end-to-end design approaches, which resulted in sub-optimal system designs. The chapter has ended with a discussion highlighting the potential of a high level of integration towards achieving more human-like tactile sensing for future next-generation tactile systems.

In this chapter, we will discuss end-to-end design strategies and methodologies to implement such next-generation fully-integrated e-skin solutions. First, **we**

M. D. Alea, G. Gielen, *Chips That Feel*, Analog Circuits and Signal Processing,
https://doi.org/10.1007/978-3-032-26584-5_3

will expound on our pragmatic technology-aware neuromorphic approach (Sect. 3.1), highlighting our approach to not blindly copy biological concepts but to only borrow techniques that help to address current bottlenecks. Next, we discuss our bioinspired approach to encode tactile signals and highlight its key advantage in improving the signal encoding fidelity while achieving a good temporal compression of the tactile data. We then show that our choice to use a spike-based approach lends well to using event-based processing such as SNNs. Such choice of bioinspired processing has been shown to ease readout demands, reflecting biological observations, resulting in power- and area-efficient readout designs. **We then show important novel functions important in next-generation fully-integrated e-skins: high-spatial-resolution texture sensing and slip detection enabled by key breakthroughs in the miniaturization and integration of sensors**(Sect. 3.2), and **high-density per-taxel readout** (Sect. 3.3). Finally, we then introduce our multi-modal and multi-scale e-skin: the INTUITIVE and DERMIS approaches, which will be elaborated in Chaps. 4 and 5.

3.1 A Pragmatic Recipe: Technology-aware Neuromorphic Design

A common criticism of the whole paradigm of neuromorphic, bioinspired, or biomimicked designs, is the perceived *stubbornness* towards strictly emulating biological concepts. This criticism is rooted in the simple fact (which we have touched upon earlier) that biological solutions are constrained by a different set of limitations and are *implemented* on a soft, organic, and unreliable substrate. Thus, *blindly* following this design principle could unnecessarily result in inefficient, unscalable, and incrementally improving designs in a significantly different substrate. A topic wherein this neuromorphic debate has largely been magnified is the topic of neuromorphic processors. Those who follow proven engineering-guided optimization principles, which we call the *neuro-skeptics*, argue that much of the improvements in today's deep learning accelerator designs, in terms of power efficiency and accuarcy, are due to the optimizations rooted in the inherent limitations of deep nanometer CMOS. Yet nevertheless, these people also often claim that world-leading designs in today's processors (i.e., NVIDIA) are still bioinspired to some extent. This is particularly true as today's compute cores still largely utilize the efficient basic paradigm of *multiply-and-accumulate*, which traces its roots in the perceptron theory (Rosenblatt, 1958), which in turn is inspired by the synaptic and membrane mechanisms of firing neurons. An easy contrast to make between the neuro-skeptic approach versus the bioinspired designs is that the former generally relies on the observations (Pfeiffer & Pfeil, 2018) that: (a) spikes are an inefficient way to represent numbers for today's hardware and workload, and (b) SNNs have not yet shown practical learning and classification advantages over its frame-based counterparts. One of the foremost pioneers in AI, Yann Le

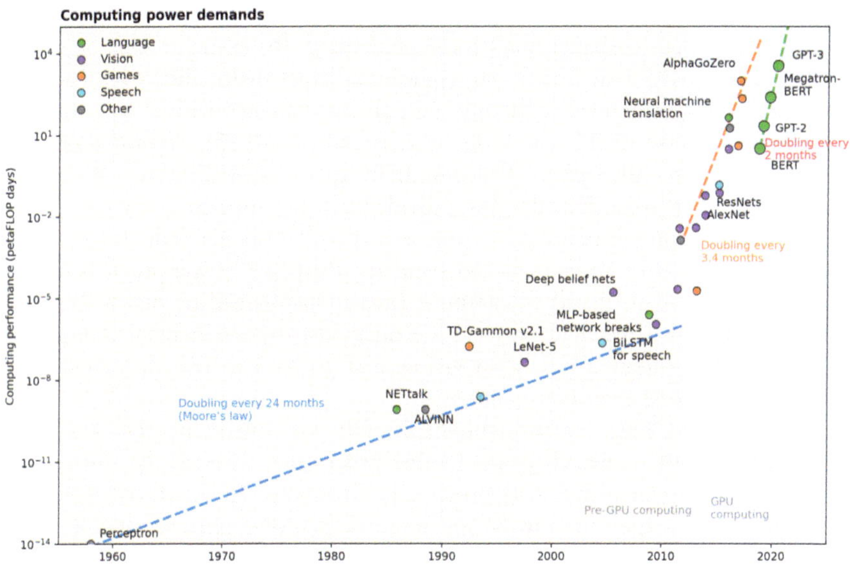

Fig. 3.1 The required amount of compute *just* to train today's models has accelerated: from doubling every 24 months (following Moore's Law), to doubling every 3–4 months, and to today's rate where it doubles every 2 months (Plot based from the dataset in Epoch AI and Our World in Data, 2025)

Cun famously said, non-verbatim, that algorithms should come first before building the hardware. "*Driving the design of such systems through hardware, hoping that someone will come up with an algorithm that will use this hardware, is probably not a good idea*," Le Cun said (Ward-Foxton, 2020).

Neuromorphs (those who subscribe to the neuromorphic principles), on the other hand, argue that neuromorphic design is beyond multiply-and-accumulte. Thus, the argument is that the apparent disadvantage of today's SNNs is that today's processors do not completely take advantage of the complexity of the brain's functions. Clearly, today's leading compute lags far behind the capabilities of the human brain in terms of: (a) complexity and cost required for training the models, and (b) its energy inefficiency. For the first aspect of this limitation, it is to be noted that the required amount of compute *just* to train today's AI/large language models have accelerated: from doubling every 24 months (following Moore's Law), to doubling every 3–4 months, and to today where it doubles every 2 months (Fig. 3.1). Meanwhile, a succinct illustration of the energy inefficiency of today's models is the visualization of a future wherein you have a "librarian in your pocket". That is, intelligence in your pocket is real-time, personalized, and secure while maintaining energy efficiency. Today's models, however, rely on mainframes to run the same intelligence. Thus, an illustration of this is that today's smartphones would need to have a 15× increase in their processor speed, and a 170× fold increase in battery capacity just to run an outdated GPT-3 model. Implementation-wise, there are three major arguments in favor of the neuromorphic design approach: (a) energy and efficiency due to sparsity and local processing, (b) asynchronous operation having near-zero dynamic power and requiring no clock distribution overhead, and (c) mixed-signal primitives emulating multiply-accumulates with less energy per operation for sparse workloads. In his article (Indiveri, 2025), *"Neuromorphic is dead. Long live neuromorphic"* Giacomo Indiveri from ETH Zurich, realizing the current state of neuromorphic designs, provided a rebuke towards going back to the foundations of neuromorphic engineering: analog subthreshold circuit design which emulates the dynamics of neurons and synapses. Despite a slower progress overall compared to its frame-based counterparts (those championed by neuro-skeptics), Indiveri argued that the field must not be misguided by incremental changes on AI benchmarks, follow a long-term vision, and go back to the fundamentals of *"understanding [the brain] by building"*.

It is important, however, to highlight the following: neuromorphic sensing is fundamentally different from neuromorphic processing, whereas the former is a topic that best resembles the work presented in this book. However, one can apply the same neuromorphic debate on the topic of neuromorphic sensing. Is there something in biology that we can emulate to improve in existing electronic skins and provide a scalable path towards next-generation e-skins? Or is it better to implement everything with tried-and-tested engineering-based strategies and chase well-known figures-of-merit? Perhaps, one can summarize this whole debate into the well-known tale of the *aeromorphic* concept: airplane versus the bird (Fig. 3.2), which perfectly summarizes the pitfalls of blindly-following neuromorphic design principles. Inspired by the flapping of a bird's feather, Leonardo da Vinci detailed ornithopther sketches wherein flight is possible via emulation with human muscles and artificial wings. Later designs by Clément Ader called the Eole utilized the era's steam-powered machines to flap artificial bat wings. Clearly, given today's jet engine technology, this is not the optimum solution for flight. As evident with modern airplanes (which do not flap their wings), the solution to emulate flight (or biology in general) is by abstracting the core principles on why it works (wings, thrust, etc.) and implement them with engineering-friendly mechanisms (turbines) on proven substrates (metallurgy). This is a core principle that we are following in this book, which we call a *pragmatic recipe* to neuromorphic design: **copy what is useful and beneficial for the specific problem, but not everything. Thus, a *pick-and-choose* approach perfectly summarizes how different substrates (nature versus solid-state electronics) result in different solutions.**

3.1.1 *Bioinspired Signal Encoding: Balancing System Power-efficiency and Signal Fidelity*

To find useful components of how the human somatosensory system encodes tactile information and adopts them into e-skin design, one must understand its behavior at

Fig. 3.2 The aeromorphic concept (as termed by Bill Dally) perfectly summarizes the pitfalls of blindly-following neuromorphic design principles. Inspired by the flapping of a bird's feathers, Leandardo da Vinci detailed ornithopther sketches wherein flight is possible via emulation with human muscles and artificial wings. Later designs by Clément Ader called the Eole utilized the era's steam-powered machines to flap artificial bat wings. Clearly, given today's jet engine technology, this is not the optimum solution for flight. As evident with modern airplanes (which do not flap their wings), the solution to emulate flight (or biology in general) is by abstracting the core principles on why it works (wings, thrust, etc.) and implement them with engineering-friendly mechanisms (turbines) on proven substrates (metallurgy)

its basic sense. Much of the neuron's computational behavior is related to its morphology. Biological neurons can be deconstructed into three major parts (Fig. 3.3a): (a) the dendrites, (b) the soma, and (c) the axon. The dendrites are branch-like structures that receive signals from other neurons (i.e., weighted input). The soma integrates the current inputs from the dendrites after low-pass filtering (i.e., the leaky capacitor). Finally, at the axon hillock, the neuron "decides" whether the integrated voltage is large enough to generate an output spike (i.e., the spike generator). The spike outputs then enter succeeding neurons through their dendrites, propagating the information encoded in the spike trains. This computational property is mimicked in artificial neural systems, with a single computational unit referred to as the *perceptron* (Rosenblatt, 1958) (Fig. 3.3b).

A fundamental theory on how neurons compute hinges on the observation derived from the Perceptron Theory (Rosenblatt, 1958) that neurons with weighted inputs and adjustable synaptic strengths can *learn* to classify patterns. Alternatively, neurons can be perceived as logic cells, where networks of neurons can *compute* any Boolean logic function (McCulloch & Pitts, 1943). Furthermore, neurons are

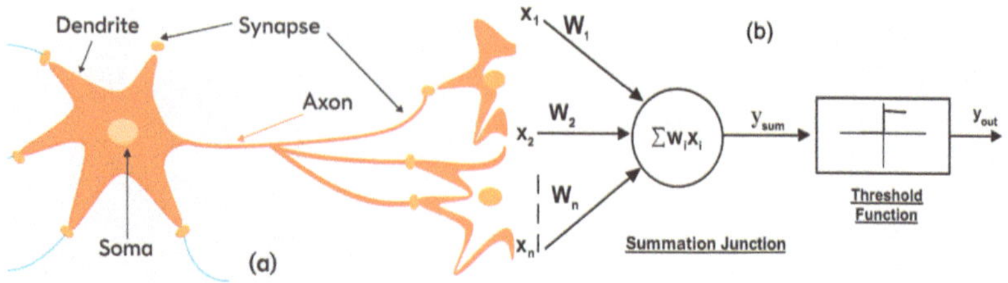

Fig. 3.3 (**a**) A biological neuron's computational operation and behavior is mimicked in (**b**) artificial neural systems

not static computers, as they also demonstrate continual learning through spike-dependent plasticity, commonly described with the phrase *"cells that fire together, wire together"* (Hebb, 1949).

How can we then unify these computational properties to the apparent lossy signal encoding happening in networks of biological neurons? Mechanoreceptors encode tactile stimuli into a continuous-time signal (i.e., a current) (Lesniak et al., 2014). This signal is then encoded into discrete-time spike trains by primary afferents (Gerstner et al., 2014) resulting in the information being encoded in an event-based manner. That is, the neuron only *fires* a spike once it *deems* that there has been significant activity in its inputs to elicit an output. Therefore, generating discrete-time spikes from this continuous-time signal inherently causes losses in the signal fidelity. This is because the conversion involves a form of quantization, where the rich temporal and amplitude information of the original analog signal is reduced to discrete events. Hence, the question arises: why settle with this solution? There are several likely explanations. One is that, biological synapses convert receptor output into spikes because the biological cabling cannot support long-distance transfer of information (Hodgkin & Huxley, 1952). That is, analog signals would degrade over long axons, but digital signals (i.e., spikes) can propagate faithfully.

Compared to any other upstream central neurons (i.e., in the brain), the key difference with tactile primary afferents in human skin is that their input comes from mechanoreceptors, whereas the central neurons accept spike inputs via synapses (Jörntell et al., 2014). **Thus, it is important to emphasize that this book focuses on neuromorphic sensing (i.e., inspired from primary afferent neurons), rather than neuromorphic processing (i.e., inspired from central neurons).** The former comes down to the design of power-efficient signal encoders inspired from neuronal firings, while the latter is more towards the design of hardware and algorithms that approximate the energy-efficient computations in the brain.

The sensing adaptation of converting continuous-time signals into discrete-time spikes, although rooted on optimizing signal fidelity given the cabling constraints in a biological substrate, can also be leveraged in sensing solutions on an electronic substrate, but with a completely different motivation. Although prior hypothesis has also looked at spike sparsity as a solution for energy-constrained coding and processing of sensory information (Laughlin, 2001) (i.e., less spikes, less energy), the predominant belief today is that the spike-based encoding is mainly a by-product of the limitation of biological cabling (Hodgkin & Huxley, 1952).

It is, therefore, interesting to see that biological and engineered solutions seem to have converged on the same implementation functionally. Level-crossing ADCs (LC-ADCs), motivated by early attempts to perform non-aliased sampling (Tsividis, 2010), demonstrate the same sparse event-based sampling and temporal compression in neurons. LC-ADCs (Fig. 3.4a) can be differentiated from the scheme used in DVS cameras (Lichtsteiner et al., 2008) (mentioned in Sect. 1.3), although it performs the same event-based sampling, in two ways: the obvious, being DVS cameras, accept photons as input, whereas LC-ADCs accept voltage as input. Secondly, DVS pixels prioritize area-efficient designs and are therefore generally less precise and noisier than LC-ADCs. LC-ADCs perform signal conversion with typical precisions of over 10 bits (Van Assche & Gielen, 2024; Timmermans et al., 2024). Similar to neurons, LC-ADCs perform event-based sampling by only generating an output when the input exceeds a certain pre-defined threshold (i.e., the membrane firing threshold in neurons). At its basic sense, this is performed by using two window comparators (Fig. 3.4a), one to detect a signal increase of 1 LSB and another to detect a -1 LSB change (Fig. 3.4b). The least significant bit (LSB) is, the smallest amplitude change that can be detected. Depending on the polarity of the signal change, the direction of the crossing is then recorded (UP/ON or DOWN/OFF). The signal is then reset to the common mode, which is ± 1LSB away from the thresholds. LC-ADCs inherently exceed the theoretical limit set by the amplitude quantizer resolution N given in Eq. (3.1) due to its oversampling properties.

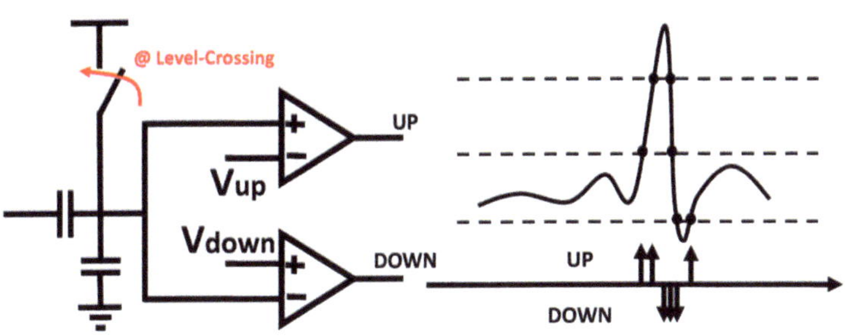

Fig. 3.4 (**a**) The level-crossing ADC (LC-ADC) performs the same event-based sparse sampling mechanism as in biological neurons. (**b**) Spike outputs are generated only when the input exceeds a certain threshold. The direction of the level crossing is then recorded via the polarity output (UP/DOWN)

$$\text{SNDR} = 6.02 \cdot N + 1.76[\text{dB}] \tag{3.1}$$

Therefore, its signal-to-noise-and-distortion ratio (SNDR), a measure of the signal encoding fidelity is determined by the timing errors associated with the samples and the precision of the re-timing interface needed to reconstruct the spike train into a synchronous continuous-time signal (Timmermans et al., 2024):

$$\text{SNDR} = -20\log_{10}(\delta \times f_{\text{input}}) - 11.2[\text{dB}] \tag{3.2}$$

where δ is the timing error and f_{input} is the signal input frequncy.

Note that this re-timing is normally performed off-chip (often in software), hence it is not included in the well-known ADC figures-of-merit. However, this re-timing interface increases the system cost due to the high frequency re-timing required to preserve the SNDR. As will be discussed later in this chapter, reconstruction/re-timing of the output time samples can be forgone through the direct use of time stamps as inputs for upstream tactile stimuli processing, such as the spiking-neural-network-based classification we performed in this book.

What is SNDR? SNDR (signal-to-noise-and-distortion ratio) is a practical measure of "how clean is the conversion?". For a sinusoidal input, it compares the power in the input tone to the power in everything else that shows up in the spectrum (within the bandwidth): random noise and harmonics (distortion),

$$\text{SNDR} = 10\log_{10}\left(\frac{P_{\text{sig}}}{P_{\text{noise}} + P_{\text{dist}}}\right) \text{ dB}.$$

It is also commonly translated into another common metric, the ENOB, computed via ENOB $\approx$ (SNDR − 1.76)/6.02.

What is Nyquist sampling? If a signal truly has no content above some bandwidth B, then sampling it uniformly at $f_s \geq 2B$ is enough (in principle) to represent it without aliasing (Nyquist, 1928; Shannon, 1949).

Why can level-crossing ADCs be "sub-Nyquist" for sparse inputs? A level-crossing ADC is event-driven: it outputs a sample only when the input crosses threshold levels. For time-sparse signals, the waveform is slow for most intervals, so its *instantaneous bandwidth* (and thus the local Nyquist rate) is low. When the signal speeds up or becomes noisy, the instantaneous bandwidth rises accordingly.

Due to the continuous-time sampling (i.e., lack of periodic/uniform sampling mechanism) nature of LC-ADCs, the converter's maximum bandwidth needs to be set such that it can *catch up* to the fastest slice of the input signal. This bandwidth is encompassed in the required loop delay (t_{loop}, Eq. (3.3)), which includes the time to (a) *detect*, and (b) *record* the level crossing, and (c) *perform* the necessary reset mechanism to set the signal level at the desired common-mode. For a sinusoidal input, the loop delay needs to be small enough to accomodate the signal's maximum slew rate (Marisa et al., 2017):

$$t_{\text{loop}} \le \frac{LSB}{A \cdot 2\pi \cdot f_{\text{input}}} \tag{3.3}$$

where f_{input} is the input signal frequncy, A is the input signal amplitude and $A \cdot 2\pi \cdot f_{\text{input}}$ is the maximum slope of the input sinusoid.

Thus, the LC-ADC effectively oversamples the input signal:

$$f_{\text{sample}} \ge 2^N \cdot \pi \cdot f_{\text{input}} \tag{3.4}$$

This event-based sampling adapts to the instantaneous bandwidth of the input signal, thus can adaptively track fast-changing signals. This adaptive-tracking property is also especially useful for time-sparse inputs, typical for tactile stimuli, as the output activity *dials down* during periods of inactivity (i.e., no contact or during rest). As discussed in Sect. 2.3, acquiring and transmitting high-throughput sensor information in the characteristically distributed manner in e-skin while maintaining energy efficiency is challenging. Thus, any form of temporal compression of the raw data should provide significant system power consumption savings. A model of this system power consumption savings will be discussed extensively at the end of Sect. 3.1.1 (*Modeling the power-efficiency advantage*).

Signal-conversion-guided solutions implemented in silicon substrates converge on a solution similar to the adaptations in biological substrates. This is despite the latter being *motivated* by a different rationale. One such aspect is the inherent temporal compression that occurs in both artificial and biological substrates. ***Level-crossing ADCs (LC-ADCs), motivated by early attempts to perform non-aliased sampling (Tsividis, 2010), demonstrate the same sparse event-based sampling as in neurons.***

The lack of samples during periods of input inactivity degrades the signal-encoding fidelity of the LC-ADC. Therefore, solutions that adopt an adaptive quantizer resolution, such as adding a triangular dither (Wang et al., 2009) or an adaptive threshold (Weltin-Wu & Tsividis, 2013; Van Assche et al., 2025), have been proposed. Similar to adaptive-threshold LC-ADCs, neurons can be thought of to encode the signal into spike trains using a similar adaptive threshold mechanism (Fig. 3.5b). ***Thus, neuronal encoders and their bioinspired equivalents (our proposed neuromorphic LC-ADCs) also get the benefit of increased signal encoding fidelity despite a slow input (where the lack of level crossings result in the lack of captured information about the signal).***

3.1.1.1 Neuromorphic Level-crossing Converters: Better Signal Encoding Fidelity

A key limitation with LC-ADC-based sampling is the lack of output spikes for constant, very slow, or very small inputs. Therefore, the lack of samples during periods of input inactivity degrades the signal-encoding fidelity of an LC-ADC. Such behavior is expected as LC-ADCs are designed to detect the signal change and not the static (i.e., DC) or slow-moving signal component. Increasing the ADC amplitude quantizer resolution N to also detect the small signal changes requires an expensive $\sim 2^{2N}$ increase in power consumption (Van Assche & Gielen, 2020). Therefore, solutions that adopt an adaptive-quantizer resolution, such as adding a triangular dither (Wang et al., 2009) or an adaptive threshold (Weltin-Wu & Tsividis, 2013; Van Assche et al., 2025) (Fig. 3.5a) have been proposed. A progressively-shrinking level-crossing window, instead of a fixed one (Fig. 3.4b), dynamically adjusts the decision level to the input signal's characteristics. Interestingly, and similar to previously discussed, biological solutions in neuronal encoding converge to a similar solution, despite being presented with a different set of constraints in a different substrate. Similar to adaptive-threshold LC-ADCs, neurons can be thought of to encode the signal into spike trains using a similar adaptive-threshold mechanism (Fig. 3.5b). In the case of neuronal encoders, the threshold is an exponentially-decaying function, not a step-wise decreasing signal. Thus, neuronal encoders and their bioinspired equivalents also get the benefit of increased signal-encoding fidelity even for slow inputs. **In this book, we borrow this adaptive resolution concept from neuronal encoding and propose to use, for the first time, a neuromorphic version of the level-crossing ADC, which we call the Neuromorphic LC-ADC (NLC-ADC).** To illustrate this adaptive-threshold mechanism in biological neurons and NLC-ADCs alike, a quick discussion on the phenomenological behavior of neurons is warranted.

In its simplest form, the neuron membrane can be modeled as a leaky integrator summing current inputs (Fig. 3.6) with a fixed firing threshold. This is commonly

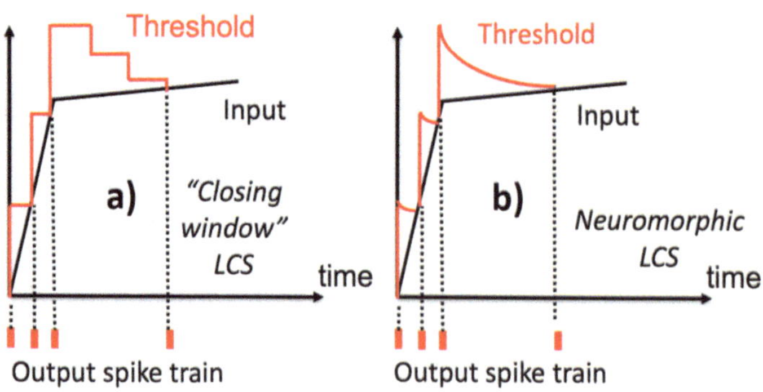

Fig. 3.5 The closing-window adaptive-resolution level-crossing sampling (LCS) in (**a**) approximates how a leaky neuron in (**b**) generates spikes

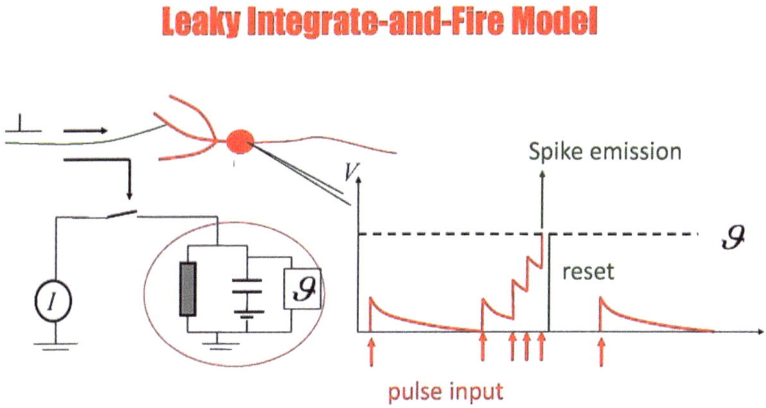

Fig. 3.6 The equivalent leaky integrate-and-fire circuit model predicts the neuron membrane and firing behaviors, despite being the least computationally expensive (Adapted from Gerstner et al., 2014, licensed under CC BY-SA 4.0)

described as the Leaky Integrate-and-Fire LIF model. However, real neuronal behavior requires models of the membrane integration and spike generation behaviors that require complex sets of differential equations. A more bio-plausible model in the Adaptive Exponential (AdEx) LIF adds an extra adaptation variable to better fit with observed firing behaviors. A plausible biophysical explanation for the observed increase is that the calcium influx during the action potential enhances the amplitude of a calcium-dependent potassium current (Gerstner et al., 2014). This adaptation variable also forms differential equations on its own, making it more computationally-expensive to predict. This adaptation variable, however, could explain the adaptive resolution-like signal-encoding property in biological neurons.

Going away from the complex differential equations formulated in the AdEx-LIF model of the neuronal spike firing mechanism, a *filter-based* approach is developed to interpret and predict neuronal encoding (Gerstner et al., 2014). In the Simplified Response Model (SRM) interpretation of neuronal encoding, the neuron is modeled as a series of linear filters with feedback (Fig. 3.7). The output spike train is passed through a leaky integrator (i.e., exponentially-decaying kernel) to form the refractory and the dynamic threshold responses embedded in two separate feedback paths. By combining the responses due to the refractory mechanism and the dynamic threshold, a similar adaptive-resolution sampling mechanism in Fig. 3.5b can be observed. This transformation is depicted in Fig. 3.8. This structure resembles a level-crossing data converter with a moving threshold derived from the low-pass-filtered output spikes. Thus, instead of a constant threshold in-between level crossings, the threshold is an exponentially-decaying value mimicking the neuron's refractory response and spike-induced membrane voltage change (Mensi et al., 2012). Note that Fig. 3.8a depicts the refractory and the dynamic threshold response appearing to have the same shapes. In reality, the two filters have different kernels, that is, both the step changes and the decay time constants are different.

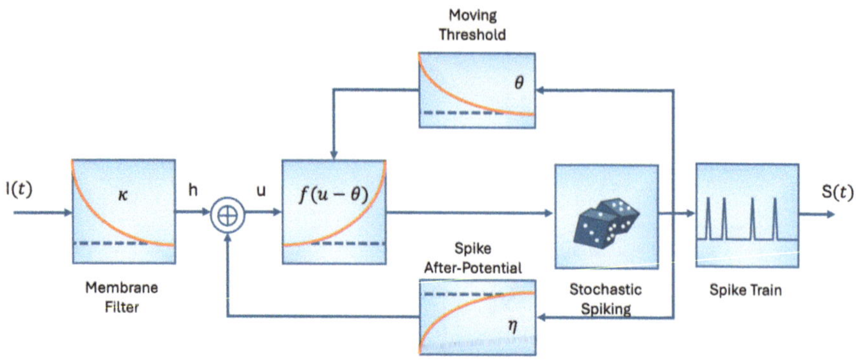

Fig. 3.7 A filter-based approach at interpreting the neuronal firing mechanism is explained in the Spike Response Model (SRM). The input spikes (impulses) are filtered with a membrane filter $\kappa(s)$. The neuron outputs a spike when the neuron membrane potential u exceeds the threshold θ. The model incorporates two feedback paths: one to set the dynamic threshold, which is the convolution of the output spike train and the moving threshold kernel. Similarly, the refractory response is added to the filtered input h, resulting in a drop in the membrane potential u, since the refractory kernel is negative by default. The refractory response is then a function of the convolution of the output spike train with the spike after-potential kernel η (Adapted from Gerstner et al., 2014)

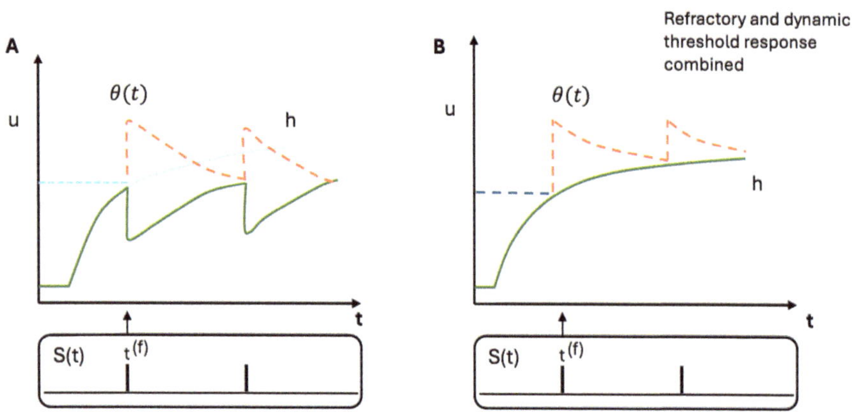

Fig. 3.8 SRM time waveform: (**a**) the evolution of the membrane potential u is reset when u hits a moving threshold $\theta(t)$. (**b**) By combining the dynamic threshold $\theta(t)$ and the reset response (i.e., the refractory response) into one single waveform, the spike trains are still consistent with the waveform in (**a**). However, such simplication of the dynamics makes it easy to visualize the SRM to be equivalent to an exponentially-decaying adaptive-resolution LC-ADC (Adapted from Gerstner et al., 2014)

The wide variety of observed firing behaviors can then be replicated by the various combinations of these step changes and decay time constants.

The key differences between the aforementioned LC-ADC and the proposed NLC-ADC are now discussed. The resulting spike outputs from both the LC-ADC and the NLC-ADC are shown in Fig. 3.9. One difference is that the NLC-ADC is inherently unipolar. That is, there is only one direction of spike, unlike in the

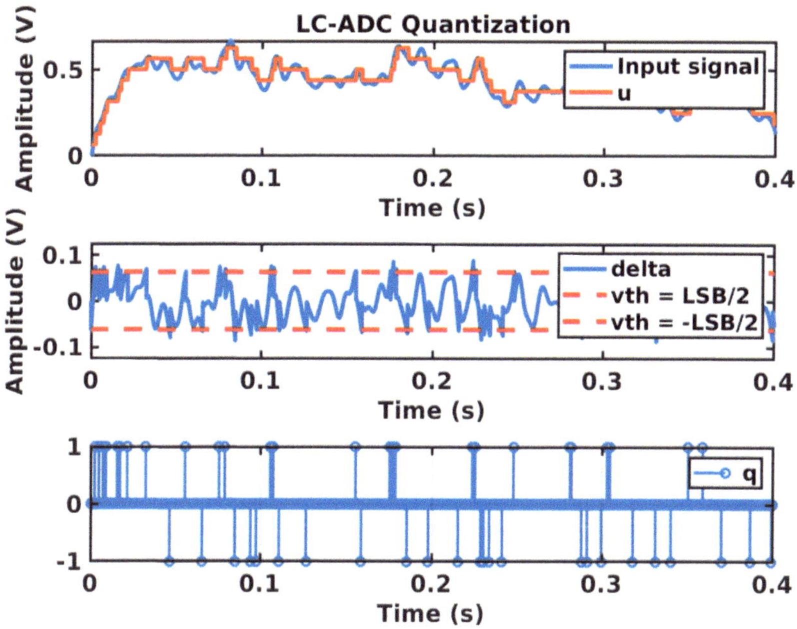

(a)

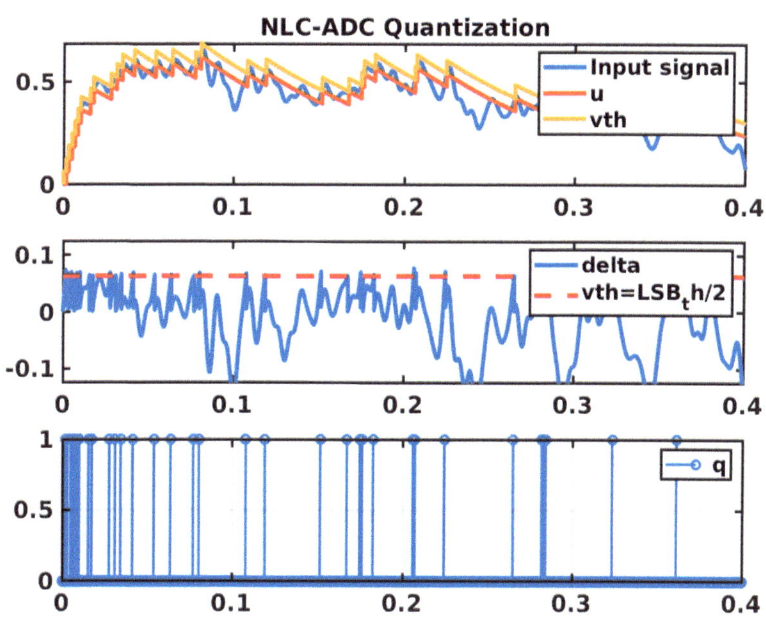

(b)

Fig. 3.9 Comparison of output spikes between conventional Level-Crossing ADCs versus Neuromorphic Level-Crossing ADCs

LC-ADC where the direction change of the level crossing is encoded into bipolar spikes (UP or DOWN; ON or OFF). Another key difference for the NLC-ADC is that it deliberately *makes* it harder for a spike to be generated right after the previous spike. This is reinforced by the adaptations in both the refractory response and the dynamic threshold, especially if the change in the sum of the refractory and the dynamic threshold is greater than the LSB. This is in clear contrast to the LC-ADC where a new spike can immediately succeed the previous one, as long as the minimum loop delay is achieved. Thus, the refractory response relaxes the loop delay requirement for the NLC-ADC. The refractory response exists in neurons as a biological adaptation to prevent the spikes (or action potentials) from propagating back towards the soma (Purves et al., 2001). Bioinspired neural networks adopt this refractory property to regulate the maximum spike frequency (Zambrano et al., 2019) and to promote sparsity in the network (Kang et al., 2025).

The adaptive-resolution spike encoding property of the SRM has motivated numerous attempts to encode signals using this approach. Much of these prior attempts are part of emerging efforts to use bioinspired asynchronous spiking neural networks for processing (Zambrano et al., 2019). Thus, encoding a continuous-time signal into spikes is necessary to provide the input. In Yoon (2017), a unipolar spike encoder based on the SRM interpretation of a neuron is conceptualized. Aside from looking at the SRM interpretation of neuronal encoding as an adaptive-resolution level-crossing ADC, the work in Yoon (2017) compares it to an asynchronous pulse sigma-delta modulator (Fig. 3.10a). Compared to a conventional sigma-delta modulator, three key differences emerge with the neuron (Fig. 3.10b): one is that conventional sigma-delta converters are clocked (typically oversampled) whereas the work in Yoon (2017) is asynchronous. Another key difference is that the integration in the feedback branch is a leaky integration. Lastly, conventional sigma-delta modulators have the integrator in the feedforward path, whereas the work in Yoon (2017) has two integrators: an input integrator before the closed-loop part and an integrator in the feedback path. Note that, when assuming that the input and the refractory (feedback) filters are the same, the transfer function in Yoon (2017) is functionally equivalent to an asynchronous version of the conventional sigma-delta modulator. It should be noted that fundamentally, the proposed neuromorphic LC-ADC resembles more an asynchronous pulse delta modulator (Fig. 3.10c) rather than an asynchronous pulse sigma-delta modulator due to the absence of the input leaky integrator. This absence of an input filter results in a spike train that is proportional to the derivative, rather than the average, preserving the essence of the event-based conversion in LC-ADCs. In the absence of the input filter, the proposed NLC-ADC accepts directly the input signal (Fig. 3.10d) . Several other bioinspired signal-to-spike encoders were developed in Zambrano and Bohte (2016). Prior works in this field, however, has almost exclusively been implemented in software.

Implementing the adaptive-resolution spike encoding principle in actual hardware requires borrowing concepts from existing analog-based neuromorphic processing work (Chicca et al., 2014), which inherently uses the same exponentially-decaying threshold as the one we need. Again, the key difference is that prior neuromorphic processing works mimic the spike-to-spike information processing in

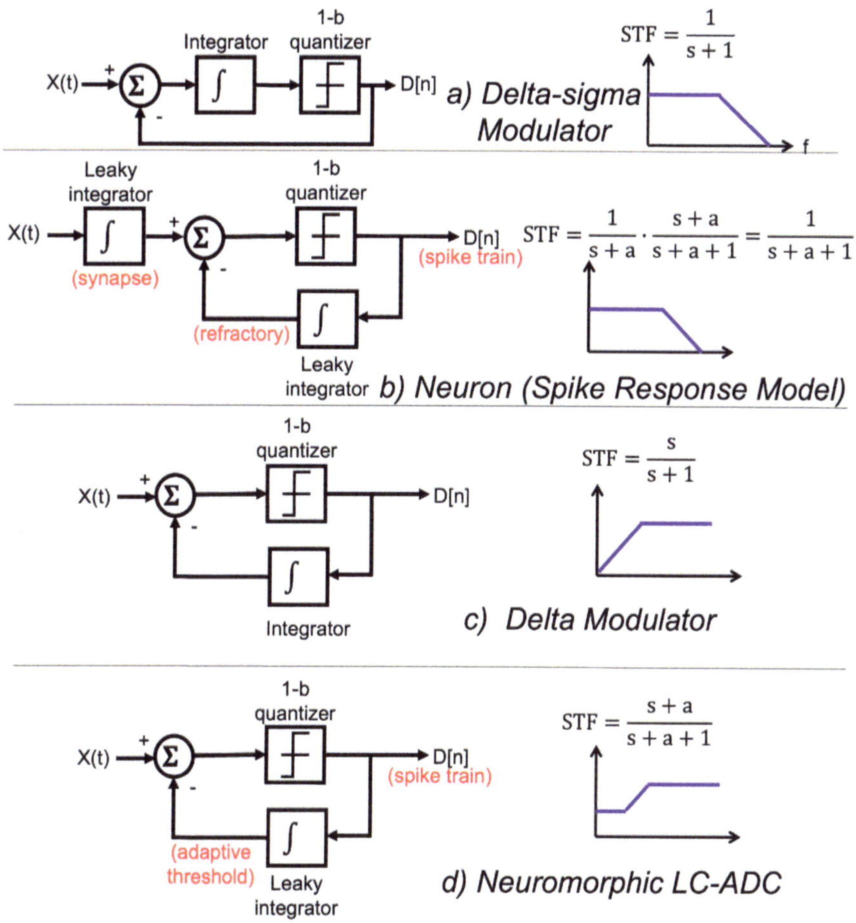

Fig. 3.10 (**a**) The delta-sigma modulator signal transfer function is similar to the neuronal encoding as depicted in the Spike Response Model (SRM) in (**b**). This results in the delta-sigma modulator output bitstream and the neuron spike train output density to be proportional to the signal amplitude. (**c**) The delta-modulator signal transfer function, on the other hand, is equivalent to the (**d**) neuromorphic LC-ADC transfer function except for the reduced attenuation at low frequencies (due to the leaky integrator). This results in the delta modulator output bitstream and the NLC-ADC spike train output density to be proportional to the derivative of the signal amplitude

the brain whereas neuromorphic sensing mimics the receptor-to-spike conversion in primary afferents. Similar to the silicon implementation of neurons (Chicca et al., 2014), the exponentially decaying kernel seen in membrane filters is implemented with log-domain subthreshold circuits. One reason is that the small subthreshold currents yield biologically plausible time constants without using very large capacitors. Secondly, the exponential dependence of the transistor current on the gate-source voltage allows easy multiplication and division of currents, which is important to establish an additive increase on the reconstruction filter output during output spikes. The effective threshold current is generated by a reconstruction filter through a subthreshold PMOS transistor. Due to the exponential dependence of the PMOS output current on the gate overdrive, a simple linearly varying gate voltage provides an exponentially decaying output current. Both the dynamic threshold and the spike after-potentials are modeled with multiple exponentially decaying kernels with different time constants, ranging from tens to hundreds of milliseconds (Mensi et al., 2012). A key disadvantage with this current-mode filter approach is that the current used as the threshold itself continually drains current away from the supply (which is signal dependent). While the current per element is typically on the order of a few tens of nanoAmpères, the aggregate consumption across a full taxel array may become non-negligible. The detailed design of such circuit will be discussed in Chap. 4.

In summary, a neuromorphic form of the event-based level-crossing ADC incorporates an adaptive threshold to improve the sampling during periods where the input is slow. Furthermore, the closing-window approach relaxes both the required level-crossing resolution and the loop delay (a larger LSB step *catches up* easier with a fast input). All this, while preserving the same temporal compression observed in the conventional fixed-threshold LC-ADC.

3.1.1.2 Modeling the Power-efficiency Advantage

In order to compare our event-based readout techniques against the frame-based counterpart in a multi-channel setting, detailed power consumption models of the following readout circuits have been implemented: *(i)* the spiking readout with regular **LC-ADC** taxels and without a closing window; *(ii)* the spiking readout with the proposed ***neuromorphic*** **LC-ADC (NLC-ADC)** taxels and *(iii)* a ***conventional*** **frame-based** readout with sensors multiplexed onto a single ADC (with a typical figure of merit of around 10 fJ per conversion step). Figure 3.11 illustrates these power models, assuming a 64-taxel array as the input of each model, which is comparable to the sensor count of state-of-the-art e-skins. The frame-based readout is modeled as a single readout channel (with multiplexing, a charge amplifier, and a SAR ADC). The power required to transmit the output off-chip is modeled as $P_{I/O}$. The spike-based readouts (LC-ADC and NLC-ADC) are modeled as multiple readout channels without multiplexing. The equivalent sampling frequency is higher than its frame-based equivalent due to its adaptive sampling nature, which needs to catch up to the maximum slope of the signal. The circuit's thermal noise (output-referred) is defined as a design parameter and directly influences the power consumption, since the noise level determines the required amplifier transconductance (assuming a constant g_m/I_D ratio). Both the frame- and spike-based readouts consider the Kylberg texture dataset (Kylberg, 2011) as the input.

Figure 3.12 shows the result of this power consumption analysis. At low signal-to-noise ratio (SNR) values (right bars in Fig. 3.12), we observe that the *multi-channel* spike-based readout with *AFE model* is more power-efficient than

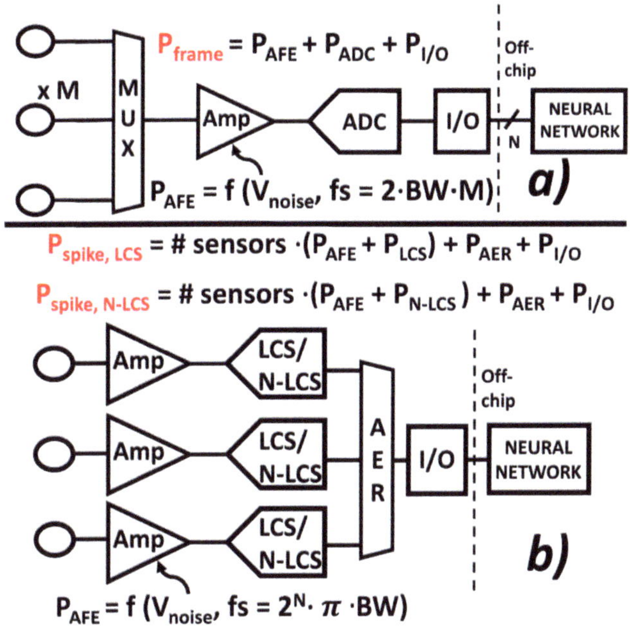

Fig. 3.11 Blocks considered for the power consumption modelling: in contrast to the frame-based readout in (**a**) where the analog front-end (AFE) circuit is shared among all sensors, the spike-based readout in (**b**) requires one AFE per taxel

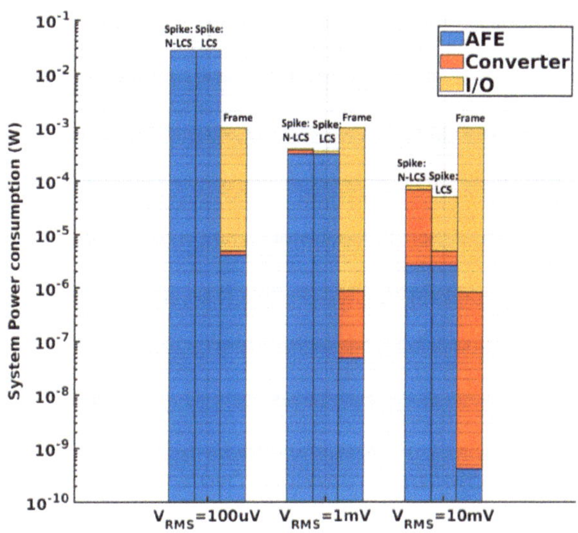

Fig. 3.12 Breakdown of the system power consumption breakdown for spike- and frame-based readouts, at three different AFE noise RMS targets. For the spike-based readouts, the conventional level-crossing converter and the neuromorphic level-crossing converter is modelled

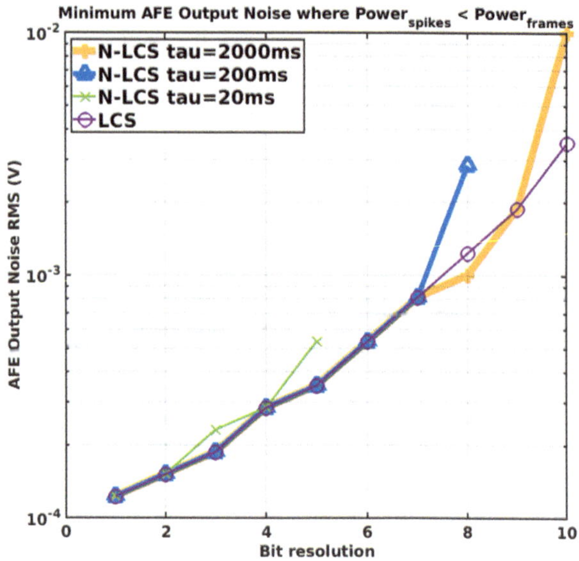

Fig. 3.13 Minimum AFE output noise above which the spiking readout is more power efficient than the frame-based one as a function of the resolution. At higher bit resolutions, the spike-based readout only becomes more power-efficient than its frame-based counterpart if the noise constraint is relaxed. The lines halting below 10 bits means that the frame-based power efficiency is better from that bit resolution onward. Plots are shown for different threshold decay constants for the neuromorphic level-crossing converter

the conventional frame-based counterpart. This result is in line with prior *single-channel* analysis, where the AFE power consumption was omitted (Van Assche & Gielen, 2020).

On the other hand, at *high* SNR values (or low noise V_{RMS} in Fig. 3.12), the power consumption of the *per-taxel* AFE in *spike-based readouts* dominates the system power consumption, making the spike-based solution less power-efficient compared to the frame-based counterpart, despite the savings in transmission power due to the *compressing* nature of spiking data. Figure 3.13 shows the allowed RMS value of the AFE output noise (cross-over point) as a function of the equivalent bit resolution. Increasing the bit resolution in both the LC-ADC and NLC-ADC restricts the minimum AFE output noise allowed to maintain the spike-based readout operating at a lower system power consumption than a frame-based readout. In other words, at higher bit resolutions, the spike-based readout only becomes more power-efficient than its frame-based counterpart if the noise constraint is relaxed.

Therefore, the analysis in Fig. 3.12 shows that the use of spike-based readouts is mainly suitable in *challenging* signal acquisition scenarios where the input SNR is low.

3.1.2 Spiking Neural Network Processing Ease Readout Demands

Our power modeling shows that spike-based readouts do have a system power consumption advantage over the frame-based counterparts. Such advantage, however, is only evident for signal acquisition scenarios wherein the allowed output-referred noise is large (Fig. 3.12). As shown for the case where V_{RMS}=10 mV, the total converter and I/O power are reduced by up to four orders of magnitude using spike-based readouts (e.g., neuromorphic LC-ADC and LC-ADC). However, per-taxel readouts do require per-taxel AFEs, instead of the ammortized readout for the frame-based approach. The power requirement for the AFE is a function of the required SNR, thus, making it less power-efficient for stringent noise requirements.

The question, thus, is if there still is a benefit then to use spike-based readout when it only is better for the specific use case of low SNR. It is therefore imperative that we verify whether we can still preserve the tactile stimuli information despite noisy frontends. For this, we look for inspiration into how biology processes the tactile information. Neurons inherently encode information into spikes with stochasticity. The random openings of mechanosensitive channels result in trial-to-trial variations in the receptor potential and spike times, resulting in spike jitter (Nierula et al., 2024). Primary afferents overcome this through population coding (i.e., combining spikes from many afferents). The inherent high-spatial-resolution of mechanoreceptors in the human skin (e.g., human fingertips), recruits hundreds of afferents when the skin is contacted by an object. Synaptic integration in the upstream neurons, thus, can *average out* any individual afferent noise to recover tactile information, such as the edge orientation (Hay & Pruszynski, 2020). Furthermore, the biological cabling to transport afferent spikes to the secondary processing in the cuneate neurons is far from perfect. Conduction velocities on this biological cabling vary broadly, resulting in further systematic spike jitter in the spike trains. However, the synaptic integration in the cuneate neurons acts as coincidence detectors, effectively correcting for the random conduction delays (Johansson & Flanagan, 2009) (Fig. 3.14).

This apparent *self-correcting*, *self-calibrating* feature of SNN-based processing thus benefits noisy signal acquisitions. To validate this, the *robustness* of tactile stimuli (*texture*) classification in the presence of modeled circuit non-idealities in the sensor-to-spike converter will now be characterized. The *Kylberg* texture image dataset (Kylberg, 2011) is used as input to a *multi-channel e*-skin sensor-to-spike converter model following the methodology in Alea et al. (2022) (Fig. 3.15). Circuit non-idealities such as the *AFE noise*, the *spiking threshold noise*, and the *spiking threshold mismatch* are added when generating the *noisy* model spikes. The tactile stimuli classification using the spiking signals generated by the model is performed by training a 256-Integrate-and-Fire-neuron, single-hidden-layer *Spiking Neural Network* (SNN), followed by spike accumulation and a *SoftMax* output layer. The network is trained with surrogate gradient backpropagation (Safa et al., 2023), then fitting a linear SVM classifier on the average spiking rate of the trained SNN's hidden layer. This choice is motivated by the facts that *i)* using a low-complexity single-hidden-layer network enables a fair analysis of the front-end hardware parameter effect, and *ii)* such resource-constrained network can be deployed directly on existing ultra-low-power SNN accelerators (He et al., 2021; Frenkel et al., 2019; Van Assche et al., 2025). Two SNN models, one trained on *clean* data (spikes generated without circuit non-idealities) and the other trained on

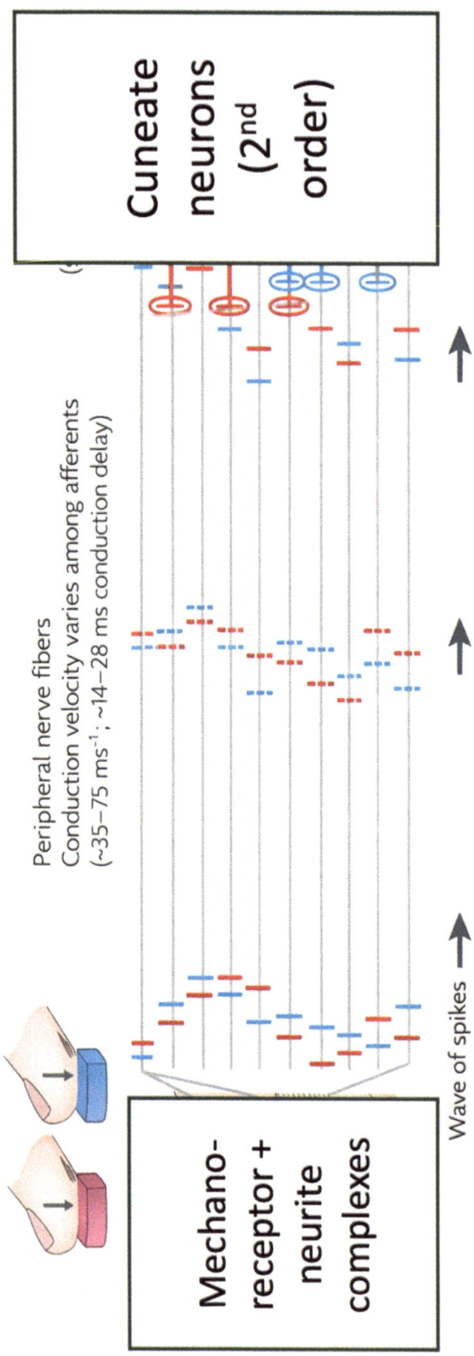

Fig. 3.14 Peripheral nerves encode touch information into a spatiotemporal spike train. Variations in the structure of these spike trains allow sufficient discrimination between different objects (e.g., blue and pink objects) at the primary afferent level. However, these spike trains then travel up to the cuneate neurons in the spinal cord via imprecise biological cabling, resulting in variable conduction delays. These variable conduction delays are taken into account and *corrected* by the cuneate neurons (Image adapted from Johansson & Flanagan, 2009)

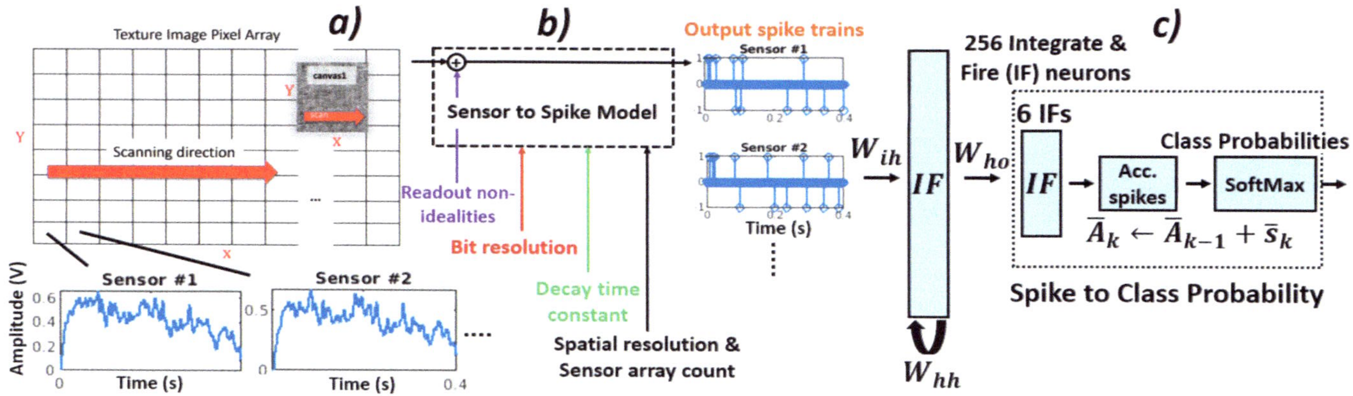

Fig. 3.15 Spiking e-skin texture classification pipeline. (**a**) Time-varying sensor signals are obtained from the texture image dataset by simulating a texture scan. (**b**) The spiking taxel conversion circuit model allows the sweeping of conversion parameters such as the bit resolution, (**c**) The spike trains are fed to a recurrent SNN through the fully-connected weight matrix W_{ih}. Recurrence is introduced by the weights W_{hh}. The output layer consists of 6 IF neurons connected through W_{ho}, the spiking output $\bar{s}_k$ of which at time step k is accumulated into $\bar{A}$. $\bar{A}$ is transformed into class probabilities using the SoftMax operator

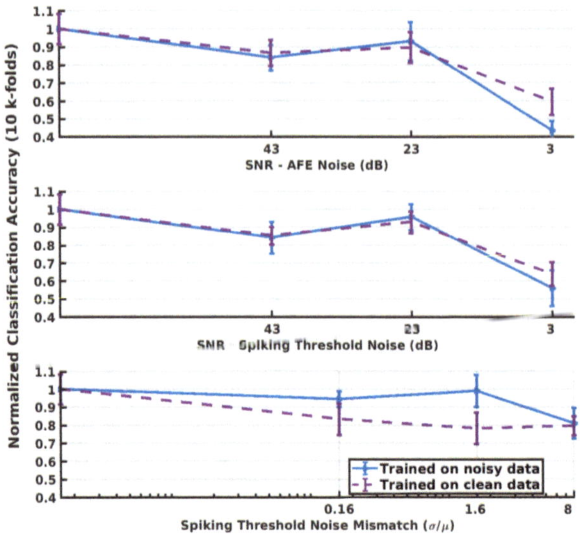

Fig. 3.16 The effect of circuit non-idealities on the classification accuracy (normalized to the accuracy when classifying *clean* data). The network is relatively insensitive (i.e., the accuracy's fluctuation is statistically insignificant) to the modeled circuit's non-idealities with a drop-off appearing only for impractically high levels of circuit non-idealities (3 dB SNR for both the AFE and spiking threshold noise; $\sigma/\mu = 8$ for the spiking threshold mismatch). The SNR values uses a signal RMS value of 135 mV, obtained from the average of the signal RMS values of texture samples across the *Kylberg* dataset (Kylberg, 2011) (scaled to fit within the AFE signal swing range)

noisy spikes, are used to classify the model spikes across various amounts of non-idealities.

The SNN training has been conducted during 500 epochs using the *Adam* optimizer (Kingma & Ba, 2014) with learning rate 10^{-3} and batch size 32. The classification accuracy and standard deviation are assessed using sixfold cross-validation (Safa et al., 2023). Figure 3.17 shows the texture classification accuracy and standard deviation as a function of the spiking data rate of our e-skin readout.

These experiments have shown that the classification accuracy (normalized to the accuracy when classifying *clean* data) is robust with respect to the individual circuit non-idealities (see Fig. 3.16). For AFE and spiking threshold SNR levels above a modest value of 23 dB, the mean classification accuracy is relatively constant as the small fluctuations in the mean value are statistically insignificant ($\sigma_{accuracy-noiseless} = 8.3\ \%$). For the dataset with modelled spiking threshold mismatch, it is shown that the other SNN model trained on *noisy* data outperforms the model trained on *clean* data. The robustness of the SNN classification accuracy to the circuit non-idealities allows relaxed conversion requirements (i.e., low SNR) for high-density *e*-skin, thus confirming the use of a per-taxel spike-based approach above a frame-based approach (see Sect. 3.1.1), as we will do in our designs of Chaps. 4 and 5.

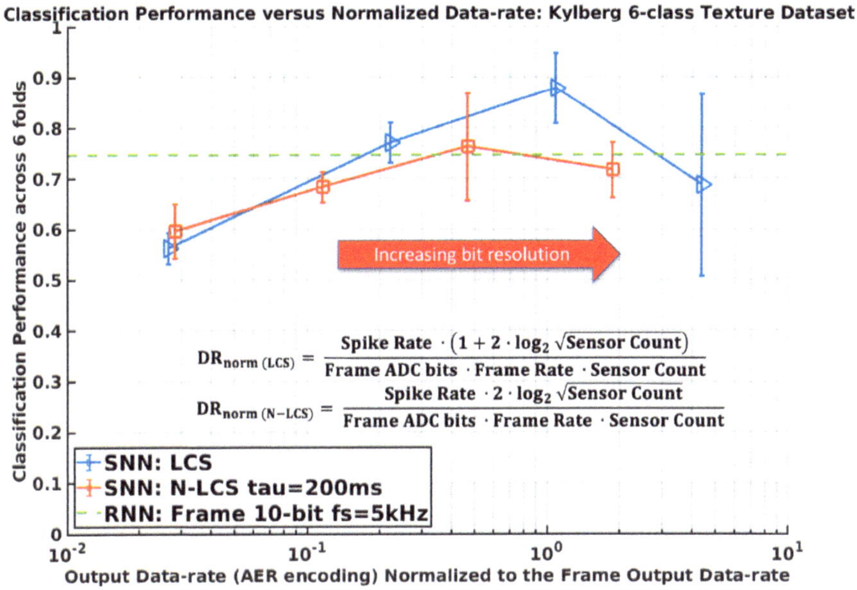

Fig. 3.17 Texture classification accuracy as a function of the e-skin readout data rate. The data rate for the spiking readout is given by the total spike rate multiplied by the row and column address bits. The LCS uses an extra bit for the spike polarity

As a reference, we also train a standard recurrent neural nework (RNN), which fits the frame-based classification of time-based tactile data, with *ReLU* neurons on the non-spiking e-skin scanning data using the same architecture as the SNN. Three interesting effects can be observed in Fig. 3.17. First, we see that the system accuracy is highly dependent on the spiking data rate, set by adjusting the bit resolution of the spike encoder, as shown in Fig. 3.15b). As expected, the accuracy increases when the data rate increases, since less signal information is lost. On the other hand, we observe a decrease in accuracy when the data rate keeps on growing. This is due to the fact that having too many spikes can jeopardize the SNN training during *Surrogate Gradient* back-propagation, leading to a decrease of the model accuracy (Safa et al., 2023). Secondly, it is interesting to note that the spike encoding and SNN pipelines (blue and orange curves in Fig. 3.17) lead to a higher accuracy compared to the baseline RNN for normalized data rates $> 10^{-1}$ in the *sub-Nyquist* region. This can be due to the sparsity in the spiking signals, which helps to increase the linear separability between data samples by providing an early compression of the sensory signals (Safa et al., 2023). Lastly, it is shown that the NLC-ADC has a slightly lower accuracy compared to the LC-ADC. It is important to note, however, that the texture classification performed here uses fast sensor signals as input whereas, the NLC-ADC' inherent advantage is for a slow input, as discussed in Sect. 3.1.1. More importantly, the network used to train the LCS spike encoder used an inherently bigger SNN model due to the $2\times$ larger input layer (to accomodate bipolar ON/OFF spikes rather than ON spikes only for NLCS). In fact, a similar comparison in Chap. 4 using output spikes from the actual chip suggests that the NLCS outperfroms the LCS in terms of the tactile stimuli classification accuracy, if the model is constrained to have the same size. Future work on the comparison of the classification accuracies between the LC-ADC and NLC-ADC using datasets with low-bandwidth sensor signals needs to be performed.

Since the system power consumption increases with the spiking data rate, we see that the *spiking* e-skin readouts (both LCS and NLCS) and SNN classifier can lead to an overall 1 to 2 orders of magnitude lower power consumption compared to the conventional RNN-based setup, while achieving a higher texture classification accuracy for normalized data rates between 10^{-1} and 1. A top accuracy of 87.92% is observed using the *LC-ADC* signal-to-spike converter, compared to 74.58% for the frame-based readout.

In summary, we have provided a breakdown of our *pragmatic* approach in adopting neuromorphic design: **copy what is useful and beneficial for the specific problem, but not everything. We have discussed that spike-based encoding not only closely mimics the biological neuronal encoding, but it also provides signal encoding fidelity and temporal compression advantages. Furthermore, spike-based encoding favorably fits event-based SNN-based classification of the spike output. Such choice of processing allows for a learning-based processing, allowing for self-correction and self-calibration-like behaviors, which relaxes the readout specifications. This results in an overall reduction in the readout power consumption by one to two orders of magnitude. We have also described how our proposed NLC-ADC architecture, which is biologically-inspired, closely resembles that of an adaptive-threshold LC-ADC, borrowing its signal encoding fidelity benefits. Thus, a *pick-and-choose* approach exemplifies how implementations across distinct substrates, such as biological and solid-state systems, can draw on shared design principles without fully mirroring one another.**

3.2 Sensor Miniaturization and Integration

A key breakthrough required in achieving more human-like tactile sensors in next-generation systems is the miniaturization and integration of high-density tactile sensors on dense electronic readouts in commercial technologies. By utilizing advanced photolithography techniques, sensors approximating the spatial resolution and sensitivity of the skin's mechanoreceptors can be achieved (Corniani & Saal, 2020). Furthermore, integrating a dense tactile sensor array on a custom CMOS or TFT readout chip, allows for a dense and power-efficient readout with local processing capabilities. The sensor-to-readout chip integration supports dense per-taxel connections, enabling *true* per-taxel sparse-sampling conversion (i.e., spikes) as performed in human afferents, while also eliminating the mechanically-fragile sensor-to-readout wiring. In this section, we will discuss the challenges and breakthroughs in depositing high-spatial-resolution tactile sensors on dense electronic substrates (Si-CMOS and TFT).

3.2.1 *High-density Piezoelectric Sensors for Texture Sensing*

Sensor-to-readout integration results in: (a) improved robustness to wear and tear, (b) reduced parasitics and noise coupling, and (c) enabling true per-taxel sensing. Piezoelectric sensor integration, owing to its simplistic structure and fabrication, offers a promising solution towards sensor integration on CMOS substrates. Prior sensor integration, however, had low spatial resolution and was integrated on coarse CMOS processes (Dahiya et al., 2011). To address these challenges, we have integrated piezoelectric sensors on commercial CMOS wafers. However, commercial foundries typically do not deposit piezoelectric layers at the fabrication level. Thus, sensor integration needs to be a post-fabrication process.

3.2.1.1 Modeling the Required Sensor Density

Human skins are innervated by Merkel cells with a fine equivalent resolution of 0.4 mm (Hay & Pruszynski, 2020). The high density of the mechanoreceptors allows the detection of fine tactile stimuli such as texture. In line with our pragmatic technology-aware bioinspired approach, we wanted to model and verify whether there are actual tactile stimuli encoding accuracy benefits into approximating the high spatial resolution of the mechanoreceptors in the fingertips. To the author's knowledge, however, modeling the effect of both the sensor count and the pitch on the texture classification accuracy has not been done previously. In this book, we have performed this search using a similar methodology as used in Sect. 3.1.2.

Our image-to-texture emulation method allows both down- and up-sampling of a subset of the Kylberg dataset (Kylberg, 2011) (class=6, trials per class=40) with emulated sensor pitch and count. Intuitively, a smaller sensing pitch allows the detection of finer details, as compared to a sensor with coarse spatial resolution, as depicted in Fig. 3.18. The larger sensor pitch can be viewed as performing a spatial low-pass filterting, removing finer details that would have been captured by a smaller sensor with finer pitch.

This modeling validates our bioinspired approach (Fig. 3.19), clearly showing the benefits of matching the human sensing spatial resolution to the texture sensing recognition accuracy, which tapers off between 0.2- to 0.4-mm pitch closely matching the biological equivalents. Meanwhile, the modeling of the effect of the sensor count on the texture classification accuracy reveals that the accuracy increases with the sensor count, but tapers off at around 196 sensors (Fig. 3.20). The INTUITIVE chip in Chap. 4 incorporates a beyond-human-level 0.2-mm sensor pitch with 192 taxels.

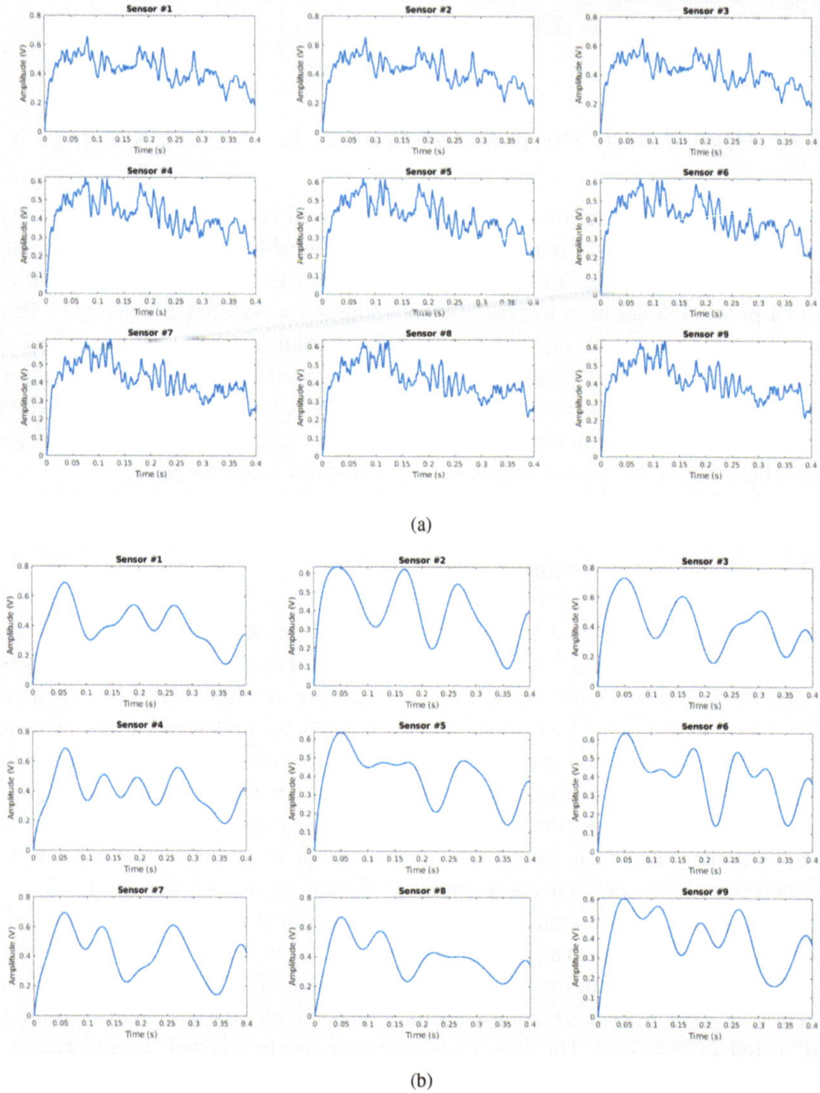

Fig. 3.18 Sensed texture signal during an emulated texture scan for varying sensor spatial resolution: (**a**) 25-μm, and (**b**) 2-mm pitches

3.2.1.2 Piezoelectric Sensor Scaling

Polyvinylidene fluoride (PVDF) is a popular dielectric choice, especially for tactile sensors, due to: (a) its high piezoelectric coefficient, (b) its bendability and stretchability, and (c) because it can be deposited without high temperatures. To achieve the spatial resolution of the mechanoreceptors in human fingertips, current

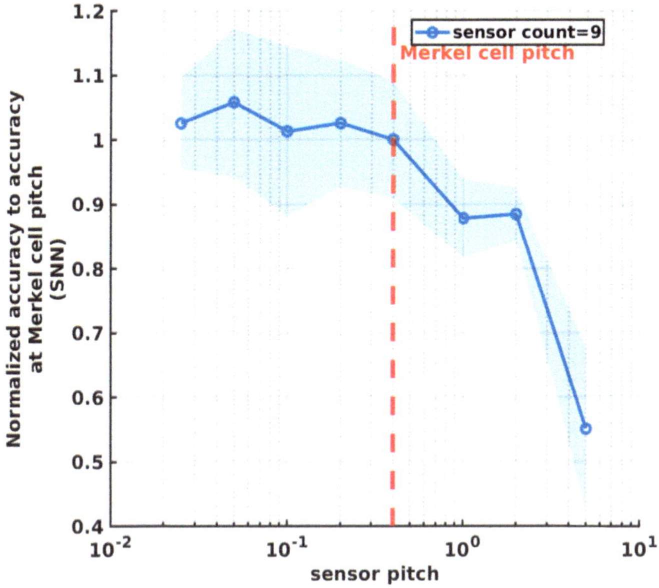

Fig. 3.19 The sweep of the texture classification accuracy, normalized to the accuracy at a 0.4-mm sensor pitch (i.e., the human equivalent), with increasing spatial resolution, revealing a higher accuracy at smaller pitches, with gains tapering off between 0.2 and 0.4 mm. The INTUITIVE chip in Chap. 4 implements a beyond-human pitch of 0.2 mm

piezoelectric arrays (Fig. 3.21) need to scale by 10× in terms of the sensor size (Halata, 1975), and 7.5× in terms of the pitch (Hay & Pruszynski, 2020). The induced charge at the sensor is a function of the dielectric capacitance, C_p:

$$\mathrm{Q_{in}} = C_p \times V_{in} \tag{3.5}$$

$$\mathrm{C_p} \propto Area_{sensor}/thickness_{dielectric} \tag{3.6}$$

Thus, to have a sufficiently large induced charge, C_p still needs to be large enough, despite the smaller sensor area. To achieve this, the PVDF dielectric thickness is reduced 10× from prior work (Dahiya et al., 2014) to 250 nm.

Aside from the increased spatial sampling fidelity, the scaling of the piezoelectric sensor dimensions also brings about two key advantages: (a) a lower input capacitance required to convert the charge into a voltage ($C_f = Q_{in}/V_{out}$), (b) a lower poling voltage due to the thin dielectric (100 V per μm PVDF thickness). Thus, the lower input-capacitance requirement allows the reduction of the per-taxel area, while the lower poling voltage reduces the risk of dielectric damage to the CMOS chip.

The fabrication cost is reduced by reducing the number of extra masks to one. Typical piezoelectric sensor structures have three layers: the bottom electrode, the

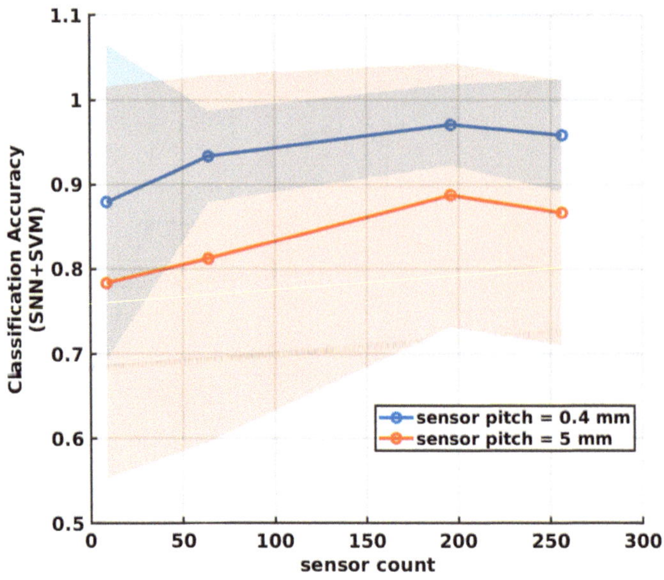

Fig. 3.20 The sweep of the texture classification accuracy with increasing sensor count, revealing that the accuracy increases with the sensor count, with gains tapering off around 200 sensors. The INTUITIVE chip in Chap. 4 implements a 192-taxel array

dielectric, and the top electrode. Since the bottom electrode also is the chip top metal, an extra mask is no longer needed. Meanwhile, patterning the PVDF layer is performed together with the top metal, sharing the same mask. Thus, the PVDF layer and the top electrode share the same patterns.

3.2.1.3 Connecting the Top Electrode

A key challenge is to connect the deposited structure's top electrode. This is because the sensor's top electrode needs to be connected to a bias voltage, which is provided by the chip. This voltage bias needs to match the voltage at the sensor's bottom electrode (typically set by the charge amplifier via feedback) to have 0V DC bias across the sensor. This is to avoid that the piezoelectric layer deforms due to the converse piezoelectric effect. Connecting the top electrode via a bondwire to a chip pad, would expose the said bondwire to mechanical contact, reducing its mechanical robustness. Thus, a clever approach is by extending the top metal to the chip edges (Fig. 3.22). The chip edges then have large pad openings that allow simple connection between the top metal and the chip (Fig. 3.23). The PCB is manufactured to have a cavity that matches the thickness of the CMOS chip, thus making the top surfaces of the chip and the PCB at the same level. This effectively reduces the length of the bondwires, which helps in increasing its robustness to wear and tear (after depositing a layer of PDMS protection, as shown in Fig. 3.24).

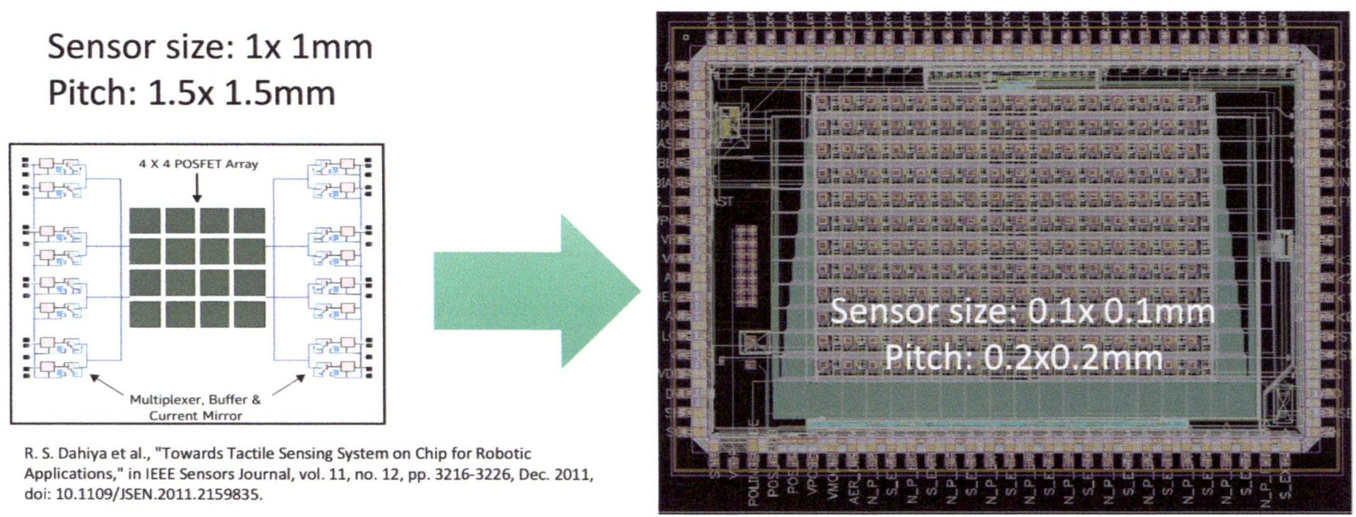

Fig. 3.21 To achieve the spatial resolution of the mechanoreceptors in the human fingertips, the piezoelectric arrays (Fig. 3.21) need to scale by 10× in terms of the sensor size, and 7.5× in terms of the pitch compared to the current solutions (Image (left) adapted from Dahiya et al., 2011)

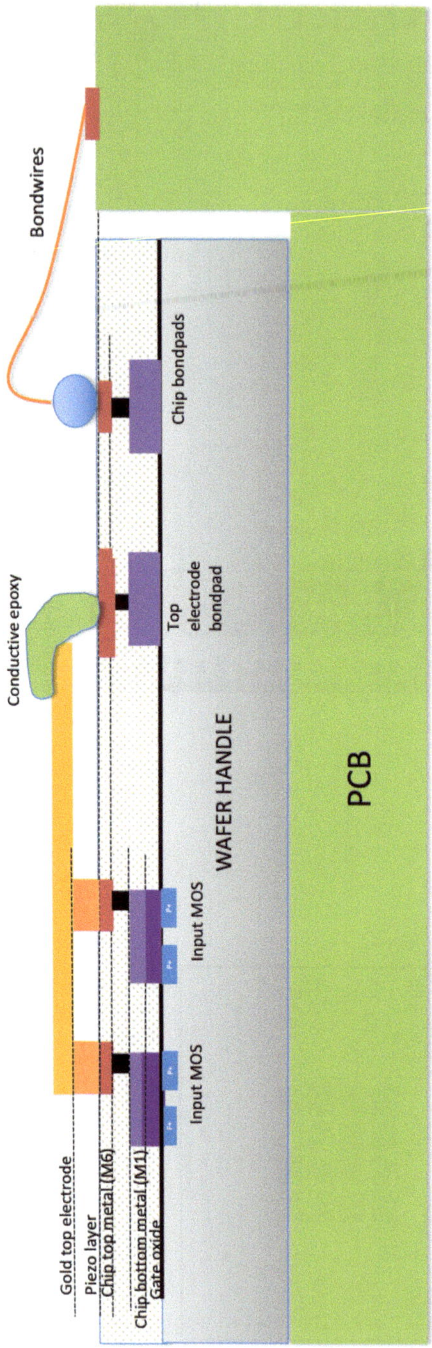

Fig. 3.22 Connecting the top electrode via a bondwire to a chip pad, would expose the said bondwire to mechanical contact, reducing its mechanical robustness. A clever approach is by extending the top metal to the chip edges. These have large pad openings that allow simple connection between the top metal and the chip. The PCB is fabricated to have a cavity that matches the thickness of the CMOS chip, thus making the top surfaces of the chip and the PCB to be at the same level. This effectively reduces the length of the bondwires, which helps in increasing their robustness to wear and tear (after depositing a layer of PDMS protection)

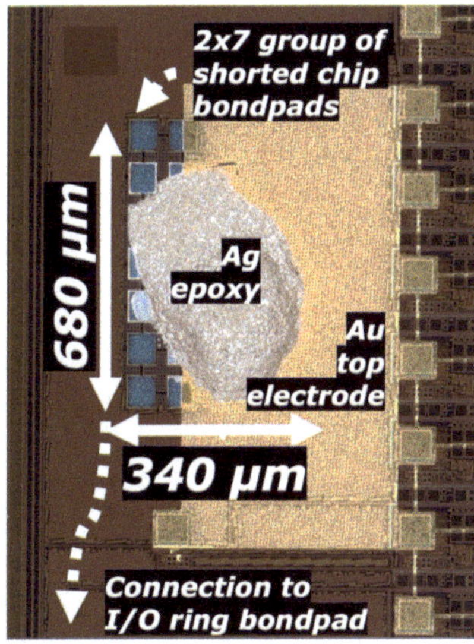

Fig. 3.23 This photo shows how the deposited sensor's top electrode is connected to the chip. A wide pad opening is realized by tiling a group of 2×7 regular bondpads. A silver (Ag) epoxy is dropped on the interface between the group of pads and the large gold (Au) portion of the deposited top electrode at the periphery of the taxel array

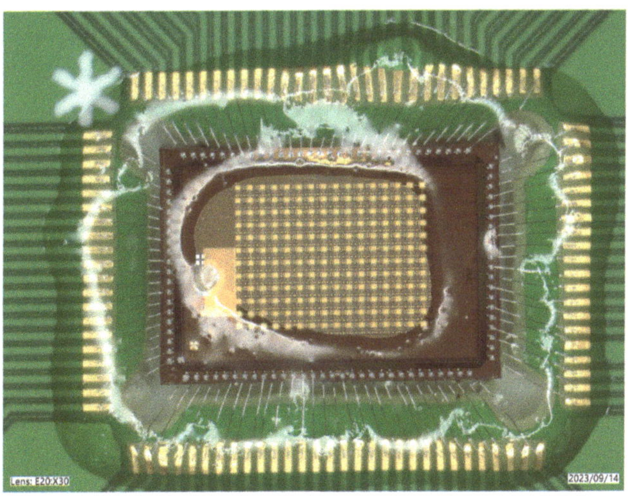

Fig. 3.24 This photo shows the sensorized Si-CMOS chip with an epoxy PDMS layer at the peripheries to protect the chip's bond wires. The PDMS is deposited in a manner that still provides an opening for the taxel array

Since the deposited top electrode needs to be connected as a whole to the top-plate bias voltage, the individual piezoelectric taxels are connected by narrow 10-μm lines to minimize crosstalk between the taxels (Fig. 3.25). Crosstalk is still possible but minimized due to the narrow PVDF layer underneath the top electrode, as they share the same mask for production.

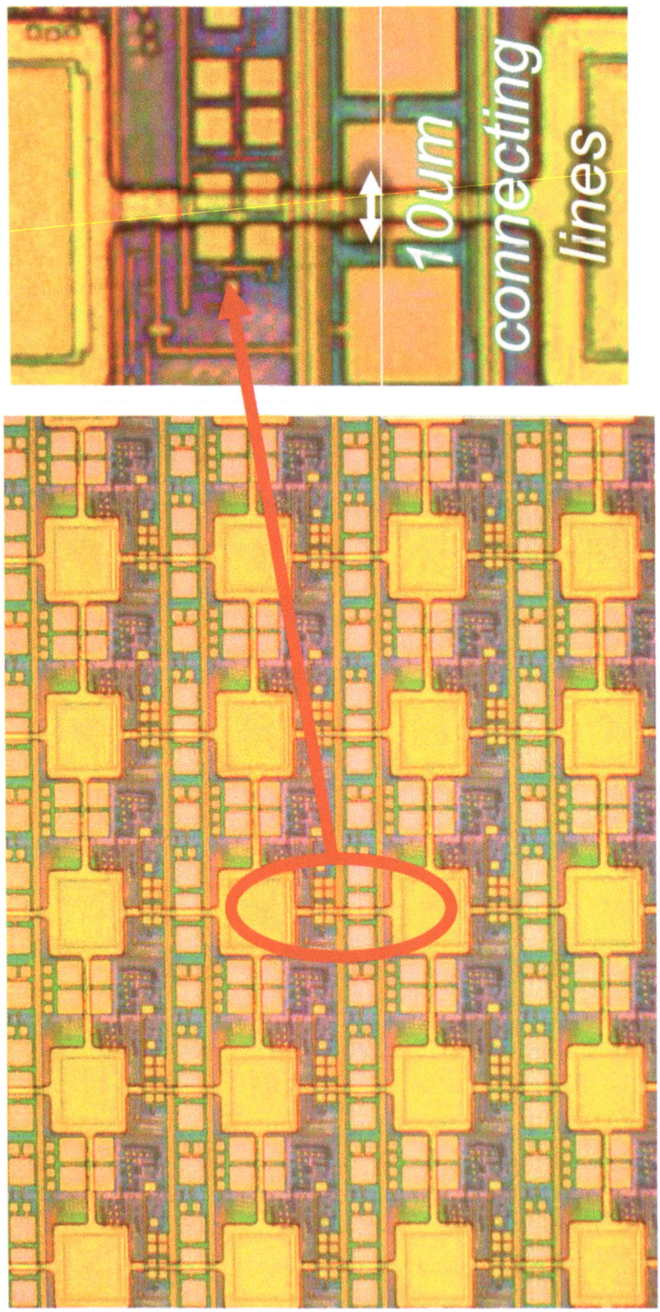

Fig. 3.25 The individual piezoelectric taxels are connected by narrow 10μm lines to minimize cross-talk between the taxels

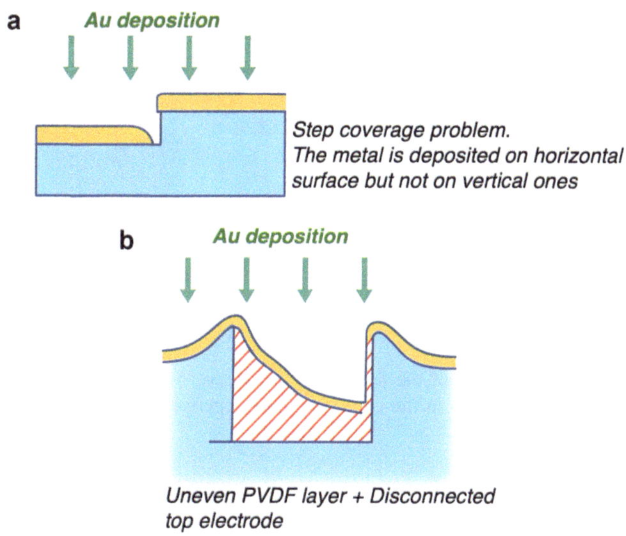

Fig. 3.26 The thick passivation layer introduces two inter-dependent effects: (**a**) non-uniformity on the deposited PVDF layer thickness, and (**b**) risk of unconnected top electrodes

3.2.1.4 Passivation Step Problem

Depositing a thin layer of PVDF introduces novel challenges, unseen in prior sensor integrations. One of them is the effect of the thick passivation layer on the CMOS chip. Due to the confidentiality of technology information relating to the thickness, we shall only refer to the thickness to be T_{pass}. The thick passivation layer would introduce two inter-dependent effects: (a) non-uniformity on the deposited PVDF layer thickness (Fig. 3.27), and (b) potential for unconnected top electrodes (Fig. 3.26).

To reduce the large passivation step, chemical-mechanical polishing (CMP) is performed on the CMOS wafers (Fig. 3.27). For the passivation layer to match the height of the chip's top metal with thickness T_{top}, CMP is performed to remove $T_{pass} - T_{top}$ of passivation layer. CMP, however, can result in over-etching, as evident in the partial removal of the chip top metal (Fig. 3.28). Figure 3.28 also shows how the height of the step is reduced along with smoothing of the passivation steps. The direction of the profilometry is shown in Fig. 3.28.

3.2.1.5 Fabrication Procedure

The deposition of the piezoelectric sensor array is performed at Fondazione Bruno Kessler, Italy. The solution has been prepared by adding 100 mg of PVDF-TrFE (Piezotech FC20) into a 3.8 mL CycloPentanone solvent. For this, a small capped glass vial (with quantity of 10 ml) has been used. The solution has been stirred using

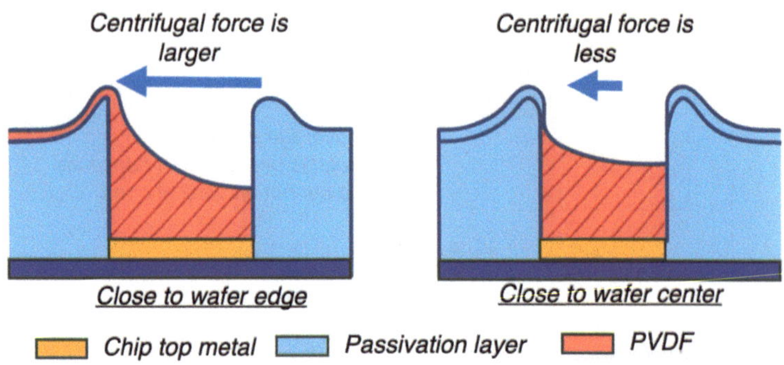

Fig. 3.27 An uneven thickness of the deposited PVDF layer is possible due to the deviation in centrifugal force between taxels in the center and in the peripheries. This is exacerbated by the large passivation step

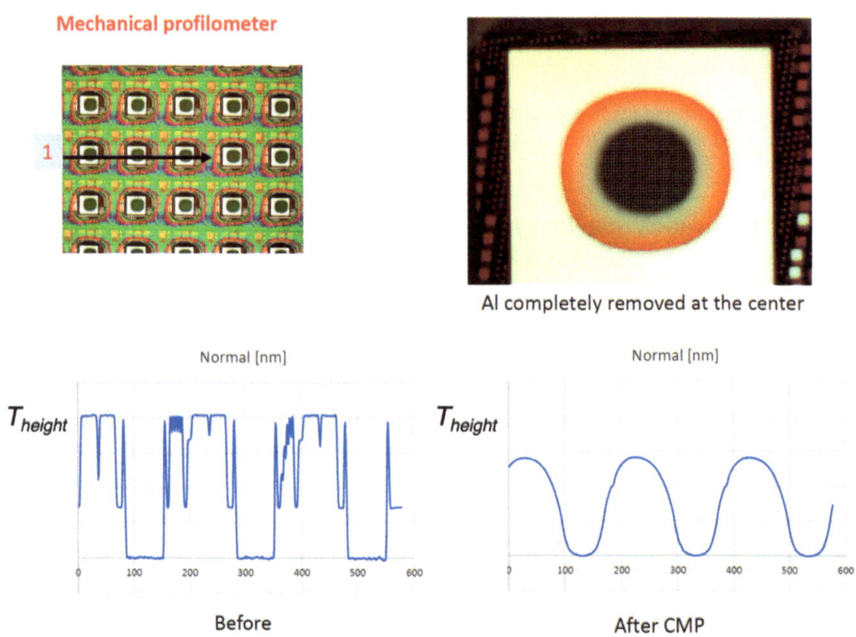

Fig. 3.28 Profilometry measurements show how the large passivation step is reduced via an extra chemical-mechanical polishing procedure done before the sensor integration

a bar magnet on a hot plate at 500 rpm at 80°C for more than 3 hours to guarantee the dissolution. The sample has been heated at 60° C on a hot plate, then it was positioned over a manual spin coater and the solution was applied over it using a micropipette (about 170 μL for 1 cm^2 sample) and spinned. The thin PVDF layer is deposited in succession. Using 2000 rpm for 60 seconds, the thickness became about 180 nm. Using 1500 rpm for 60 seconds, the thickness became about 270 nm.

After the deposition, the samples have been annealed in a vacuum oven at 140° C for 2 hours and followed by the natural cooldown.

The heating of the sample during spinning was required to obtain the desired structure of the other polymer; for PVDF only it would be possible to use it at room temperature too, but the viscosity would be higher and which would result in a thicker film.

Poling is performed to activate the piezoelectric properties by aligning the deposited dipoles on the PVDF layer. The poling waveform is a step-wise increasing sinusoid at a frequency of 0.05 Hz. During fabrication, due to the uneven thickness of the PVDF layer, shorts between the top and the bottom electrodes can form. The high-voltage poling procedure helps to eliminate these shorts by allowing a large current compliance through the sensors, effectively *burning out* any short.

3.2.2 Capacitive Shear Sensors for Slip Sensing

To match the human somatosensory system, the requirements for *next-generation* tactile sensors must be based on the low-latency capabilities of the human hands to detect slip with high spatial resolution and to detect the relevant contact parameters that depend on the current grasp state. Human skin, due to its mechanical compliance, is able to detect complex tensile, compressive, and torsional forces over large areas of the skin during object handling. The high density of receptors on the fingertips is then hypothesized (Khamis et al., 2014) to encode the moving strain wave with a spatiotemporal spike pattern towards the nervous system. Thus, slip detection requires the large-area, high-spatial-resolution sensing of shear and normal forces. However, existing shear-force-capable tactile sensors (Boutry et al., 2018; Cheng et al., 2010; Yousuf et al., 2022; Khamis et al., 2018; Ulloa et al., 2022) are bulky and not flexible; thus they cannot conform to curved surfaces. Furthermore, they lack the required spatial resolution to capture the fine-grained evolution of the shear force during object manipulation. Lastly, the existing shear sensors are interfaced to off-the-shelf electronics and are connected via cabling that are prone to noise coupling and wear and tear.

There are three major ways to detect slip: (a) optical-based tactile sensing, (b) vibration-based sensing, and (c) friction-based sensing.

In a commercial optical sensor (Yuan et al., 2015), the complex force profile when an elastic polymer contacts an object modulates the position of dots, which is captured by a high-resolution camera. Such an approach yields a high-spatial-resolution snapshot of the contact forces, allowing the use of image entropy extraction to predict slip. While this achieves excellent slip detection, such an approach results in excessively-large acquisition data rates and complicated upstream entropy extraction processing.

An alternative approach completely devoid of force sensing is detecting slip through micro-vibrations, generated when a textured object slides across the sensor. This approach often relies on simple (often single *taxel*) sensors such as a PVDF film (Romeo et al., 2017) or a force-sensitive resistor (FSR) (Damian et al., 2010). Extracting slip from the sensor output is then done through envelope detection (Romeo et al., 2017), Fast Fourier Transforms (FFT)(Fernández et al., 2014), or Discrete Wavelet Transforms (DWT) (Wang et al., 2016).

A simpler approach to slip detection is a friction-based approach (Tremblay & Cutkosky, 1993) by which the normal and tangential forces are acquired during incipient slip. This technique hinges on the fact that to prevent slip, the normal force, f_{norm}, applied by the grasp should be higher than $f_{tan} \cdot \mu_s$, where f_{tan} is the tangential force at the contact and μ_s is the static friction coefficient, known *a priori*. While, in theory, simple algorithms would suffice to detect slip, the approach in Tremblay and Cutkosky (1993) uses bulky and expensive multi-axial force/torque (F/T) sensors to detect the tangential forces.

We leverage miniaturized flexible capacitive sensors to utilize the simpler friction-based technique, therefore avoiding complex upstream algorithms for slip detection, while allowing for a small form factor and curvature-compliant sensing. To retain true per-taxel sampling, as performed in human hands (Delhaye et al., 2021), we adopt the approach used in the piezoelectric sensor integration described before, to integrate shear- and normal-force-capable capacitive tactile sensors on a large-area flexible thin-film substrate.

3.2.2.1 Achieving Human-level Slip Detection Capabilities

Drawing on the slip-sensing behavior of the Meissner and Pacinian corpuscles in the human fingertip, we derive the sensing requirements for the proposed slip-capable sensor. Slip detection likely combines the Meissner flutter cues (~5–50 Hz; pitch ~0.5 mm) with the Pacinian high-frequency vibration cues (~40–400 Hz; pitch ≈2–3 mm), and localized slip responses are common in Meissner/Merkel cells but rare in Pacinian corpuscles (Johansson & Flanagan, 2009; Corniani & Saal, 2020; Zangrandi et al., 2021). These biologically-inspired specifications and the values achieved in our designs will be summarized in Table 3.1. Detailed implementation discussions are provided in Chap. 5.

Prior studies (Delhaye et al., 2021) suggest that the fingertips are populated with high-density Meissner and Pacinian corpuscles which can detect both normal and lateral forces. During object handling (and during object slip), the high-density population of the Meissner and Pacinian corpuscles in the skin sense a complex spatiotemporally-evolving shear strain over the fingertips. Thus, two key requirements emerge: the need for (a) high-density multi-axial force sensing, and (b) large-area coverage. In this book, the fingertip is considered "large-area" because its surface area is orders of magnitude larger than typical chip-scale devices. Meissner corpuscles populate the fingertips with an estimated spatial resolution of 0.5 mm (Delhaye et al., 2021). The index fingertip has a mean area of 400 mm^2 (Johnson & Blackstone, 2007). Limited by the tangential-force dynamic range, which corresponds to the maximum unidirectional lateral displacement of the top electrode that still preserves electrode overlap in our design, the achieved shear-

Table 3.1 Sensing requirements for slip detection based on their biological equivalents

	Requirement	Achieved
Pitch	500 μm	600 μm
Bandwidth	Adaptive event-based (max: 400 Hz)	Adaptive event-based (max: 500 Hz)
Normal force dynamic range[a]	5 mN to 4 N	2 mN[b]/ 46.5 mN[c] to 37.1 N
Lateral force dynamic range[a]	5 mN to 1.4 N	2 mN[b]/ 46.5 mN[c] to 6.2 N
Force sensed	Normal and shear forces	Normal and shear forces
Detection latency	50–100 ms	Achievable in theory (based on spike rates)

[a]Theoretical dynamic range values derived in Appendix B
[b]AFE-noise-limited
[c]ADC-resolution-limited

sensor pitch is 0.6 mm (which is 3.33× finer than the prior state of the art in Su et al. (2015)). Our prototype of 2×8 taxels is implemented in a chip of 36 mm^2 area. Scaling the device area to achieve the target 400 mm^2 coverage is primarily an engineering challenge, constrained by the cost and the access to wafer-level fabrication.

In everyday-manipulation-like conditions, the forces experienced by the fingertips are characterized to be around 0.2 to 4 N for the normal force, and around 1.4 N for lateral loads (Birznieks et al., 2001). Meanwhile, stroking experiments reveal an ultra-low force sensitivity of 5 mN (Bai et al., 2015). *"Finger-jamming"* experiments provide a practical upper bound for strong normal loading, with a reported maximum of 42 N (Hohendorff, 2019). Given the sensor dimensions and the dielectric material used, our analysis in Appendix B shows a maximum normal force and lateral force of 37.1 N and 6.2 N, respectively. The minimum force detectable is limited by the level-crossing ADC, which translates to a minimum force change, before a single output spike is generated. Given a minimum ADC detectable voltage signal (after the capacitance-to-voltage frontend) of 45 mV (differential) with a full-scale signal of 6 V (differential) and a sensor full-scale force range of 6.2 N, the equivalent minimum force signal is derived to be 46.5 mN. Meanwhile, the noise-limited condition (considering only the capacitance-to-voltage noise), we get a lower 2 mN bound (as depicted by the SNR plot across sensor transduction conditions in Fig. 3.29).

A 20× larger prototype has been experimentally validated to transduce a maximum pressure of 10 kPa (78.5 N for a circular MTS Criterion indenter tip of 100-mm diameter) before its output saturates. While our original plan was to include integrated sensors in the current prototype, unforeseen constraints in the fabrication capacity prevented us from doing so. The sensing principle itself has already been experimentally validated using the larger-scale prototype sensor, we report here the results of the measurements of the readout IC. The readout has thoroughly been tested using external off-chip capacitors, which can be adjusted in real time to mimic the expected sensor signals. Together, these results indicate that the remaining step

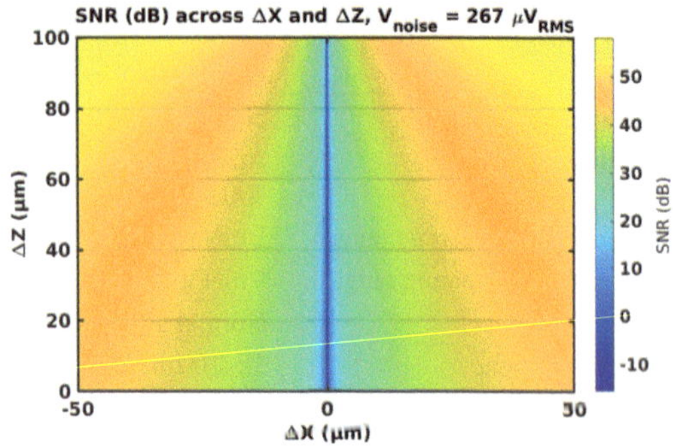

Fig. 3.29 Equivalent signal-to-noise ratio (SNR) of the DERMIS system considering the readout noise

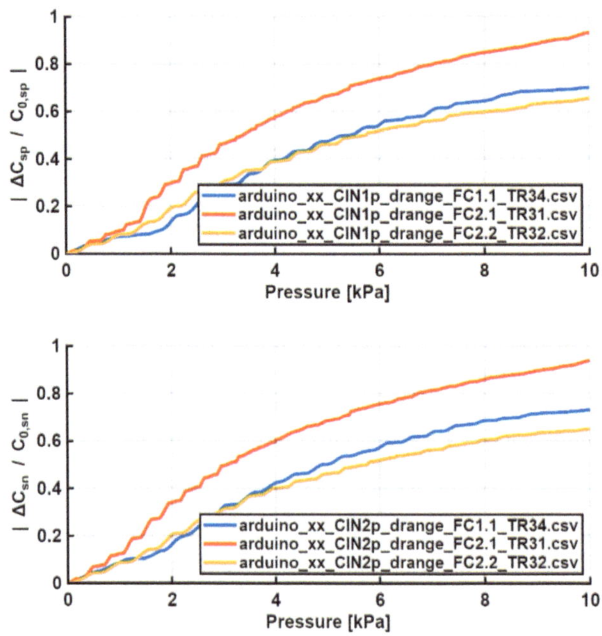

Fig. 3.30 Normalized capacitance versus pressure measurements for multiple sensor samples. C_{sp} and C_{sn} denotes the two sides of the differential capacitance

of fabricating and integrating the miniaturized sensors onto the presented chip, is primarily an engineering task, rather than a fundamental limitation of the sensing concept or the readout architecture. It is clear though that future work needs to validate experimentally the derived performance of the miniaturized version of the fabricated differential capacitive sensor prototype.

We include here measurement results (Fig. 3.30) from multiple sensor samples to reflect the spatial variation one would observe in a monolithically-fabricated sensor array. All samples were fabricated on the same wafer, so this captures the inherent within-wafer variation across devices and thus closely emulates the behavior of an arrayed sensor. Possible sources of sensor mismatch can be attributed to mismatches in: (a) the dielectric thickness, and (b) the top-to-bottom electrode alignment (due to the manual alignment procedure).

Incipient slip (partial slip) detection is typically a pre-cursor to gross slip (total loss of contact with an object). Thus, a fast (50–100 ms latency Johansson & Flanagan, 2009) and accurate detection of incipient slip is crucial to maintaining a stable object grasp by regulating the force applied by the artificial hand. While real-life grasping experiments involving slip detection were beyond the scope of this book, the high maximum spiking output rate of the DERMIS system of 100 $kevent/s/taxel$ (which exceeds the sampling rate of prior state-of-the-art fixed-sampling-rate tactile sensors), could theoretically allow fast slip inference, also considering the use of fast time-to-first-spike algorithms (Lee et al., 2017).

3.2.2.2 Design Rationale

A key limitation with prior tactile sensing, especially capacitive and piezoelectric sensors which require bottom and top electrodes, is the difficulty of physically connecting the exposed top electrode. Such a connection (commonly with wire-bonding) is prone to wear and tear due to the frequent mechanical contacts. The proposed capacitive sensor overcomes this challenge by adopting a "floating top electrode" approach. That is, the electrical connections are only done on the two bottom electrodes, while the middle electrode (see Fig. 3.31) is left floating. Thus, an equivalent series capacitance is formed. Since the two bottom electrodes are at the same substrate as the flexible readout, the top and bottom electrodes are connected at chip level. To form a differential structure, an identical capacitive structure inverted along the sensing direction (i.e., the X or Y direction) is added. The differential structure not only enhances the sensor's robustness to noise, mismatch, and parasitic capacitances, but also allows both normal and shear force measurements, as will be explained below.

Capacitive tactile sensors have been demonstrated to detect normal forces (Janotte et al., 2022; Rosa & Kraft, 2024). However, slip sensing requires that the capacitive sensors are modified to be able to measure shear forces while retaining

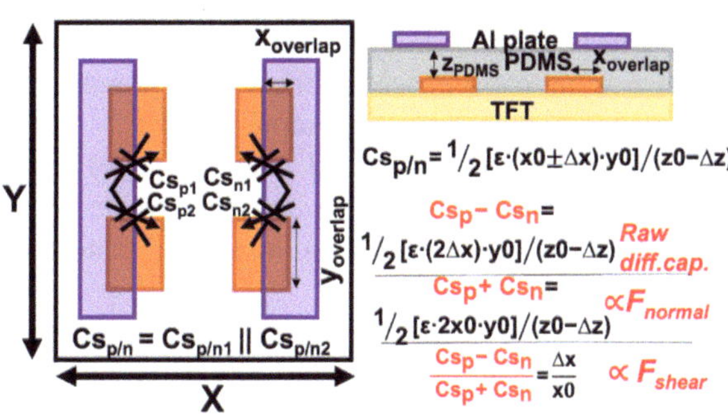

Fig. 3.31 Differential capacitive sensor for both normal and shear force detection, designed to be integrated on top of a TFT substrate with integrated readout electronics. On the right are the equations to extract the normal and shear forces

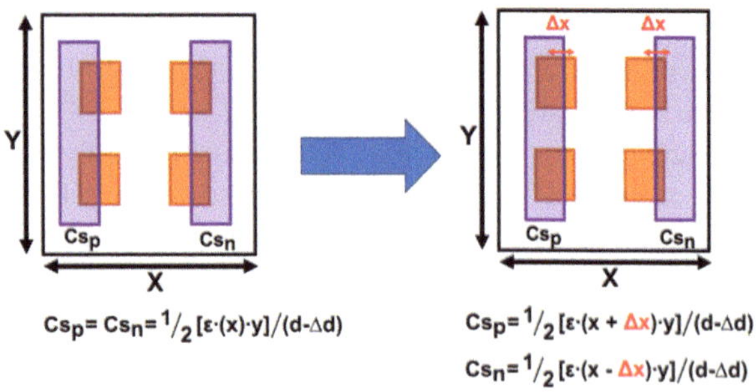

Fig. 3.32 Depiction of the differential sensitivity of the capacitive sensor to the application of shear force

the capability to detect normal forces. Furthermore, the capacitance change due to the normal force should be independent from the shear force, and vice versa. Prior related work (Cheng et al., 2010) utilizing the same floating top electrode cannot independently measure the shear and normal forces. That sensor also relies on a "bump" microstructure, which is not fabricated trivially, to be able to measure the shear forces. Furthermore, the sensor does not perform differential measurements and is thus prone to external noise coupling and parasitic capacitances.

The differential capacitive sensor output used in this work is detailed in Fig. 3.31. Shear-force sensing can be performed by changing the overlap area between two parallel plates. This results in a lateral-force-dependent capacitance change on the differential sensor, but with the opposite magnitude (i.e., one increasing, the other decreasing). This is depicted by Fig. 3.32. To make the output value independent from the normal force, which affects both sides of the differential structure, a ratiometric approach is performed. By dividing the differential capacitance by the common-mode capacitance (i.e., the sum of the two pairs of capacitances), the resulting value is independent of the normal force. Meanwhile, the output due to the normal force is inherently independent of the shear force, since it considers the common-mode capacitance, which is constant (i.e., a change in one capacitance pair results in an opposite change in the other one).

The impact of any parasitic fringe capacitance is reduced by (a) using a thin dielectric and (b) sufficiently separating the two floating top electrodes. Consequently in our design, the fringe capacitance is approximately 32× smaller than the primary overlap plate capacitance (Fig. 3.33, Eqs. (3.7) and (3.8)). Moreover, the differential capacitance-to-voltage conversion effectively transforms this fringe term, which appears as a mutual capacitance between the differential sensors, into two shunt capacitances to the virtual ground, thereby further suppressing its contribution to the differential output.

We estimate the normalized overlap (plate) capacitance and the normalized parasitic fringe capacitance as follows, where W is the plate area, H is the dielectric thickness, and S is the horizontal plate separation of the top electrodes

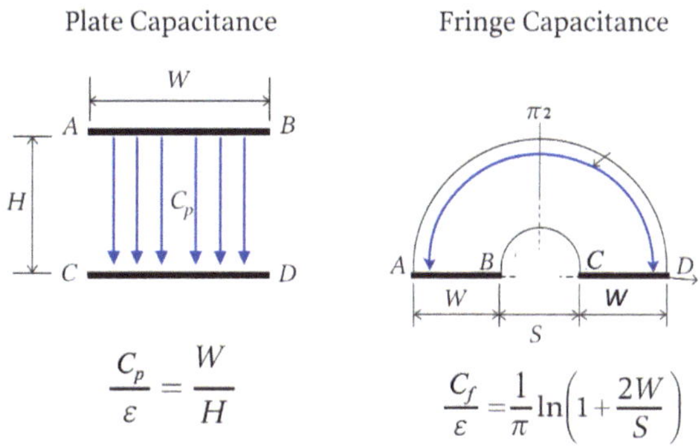

Fig. 3.33 Analytical expressions for both the primary overlap plate capacitance and the parasitic fringe capacitance of the capacitive sensor

(all dimensions in μm). The capacitance values are normalized by the absolute permittivity ε.

$$\frac{C_p}{\varepsilon}\,[\mu\text{m}] = \frac{W}{H} = \frac{125 \times 50}{125} = 50, \tag{3.7}$$

$$\frac{C_f}{\varepsilon}\,[\mu\text{m}] = \frac{1}{\pi}\ln\left(1 + \frac{2W}{S}\right) = \frac{1}{\pi}\ln\left(1 + \frac{2 \cdot 125 \cdot 50}{100}\right) \approx 1.54. \tag{3.8}$$

3.2.2.3 Fabrication Procedure

The fabrication of the capacitive sensors is performed at the Leuven Nanocentre, Belgium. The sensor fabrication follows a novel sacrificial layer approach (Rosa & Kraft, 2024) (Fig. 3.34). The process begins with thermal evaporation of a 100-nm Al sacrificial layer on a 4-inch Si wafer, followed by a silane (VM651) adhesion coat. A PDMS (SYLGARD 184, 10:1) layer is then spin-coated to 200 μm and oven-cured; a brief silane coat is repeated to address PDMS hydrophobicity. The electrodes are patterned using ma-N1420 photoresist with standard expose/develop steps; a 100-nm Cr/Au/Cr stack is sputtered, and the metal is removed by lift-off in acetone. The dielectric layer is then fabricated according to the sensor type; for a flat sensor, this consists of an adhesion layer followed by another 200 μm PDMS deposition.

In summary, we have reported our breakthroughs in sensor miniaturization and integration on both commercial Si-CMOS and TFT substrates, with the goal of achieving more human-like e-skin systems.In addition, we have derived and summarized key biologically informed system specifications to support this goal, particularly for slip sensing. For texture scanning, we have shown that increasing sensor count and spatial resolution improves texture classification accuracy, thereby informing the target values for both parameters. Also thanks to these sensors, designs to be presented in the next chapters achieve the lowest spatial resolution in both fine-texture and slip-sensing e-skin solutions. The

Fig. 3.34 The sensor fabrication follows a novel sacrificial-layer approach with 6 steps. The sensor is then designed to be inegrated onto a TFT chip via an alignment-based step after both sensor and TFT fabrication

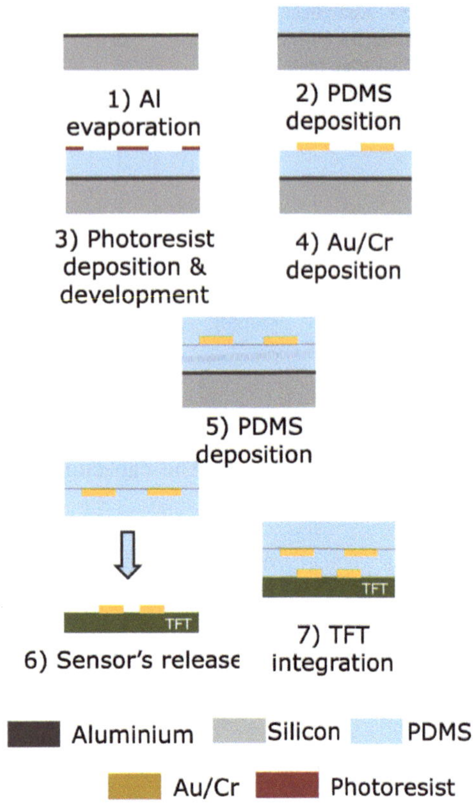

integration of these sensors on mature substrates with high-density electronics allows for power- and area-efficient per-taxel e-skin systems, a step closer towards the similarly efficient true per-taxel sampling seen in the human somatosensory system. Lastly, we have discussed engineering approaches to solve the inherent mechanical wear-and-tear issues in tactile sensor designs.

3.3 Per-taxel Readout

As discussed in Sect. 2.6, per-taxel readout offers several system-level advantages in terms of spatiotemporal signal compression, mechanical robustness, and improved sensor signal-to-noise ratio. However, the promise of per-taxel readout can only be realized if the two major limitations are addressed: the per-taxel area and power consumption challenges. In this section, we compare state-of-the-art works in prior fully-integrated e-skin solutions in terms of their per-taxel readout power consumption and area. We then see how far prior work, is trailing the mechanoreceptors and afferents in human fingertips, which sense fine localized texture and slip—two of the most challenging applications in today's e-skin systems.

3.3.1 *Area Challenge*

3.3.1.1 Localized Texture Sensing

In Fig. 3.35, we interestingly see that a lot of the prior readouts already are below the spatial resolution of Merkel cells (Hay & Pruszynski, 2020). Merkel cells form receptive fields at the human fingertips. These are formed by the random innervation of primary afferents to multiple Merkel cells (Lesniak et al., 2014). Neuroscience measurements suggest that sub-field patterns have a 0.4-mm spatial resolution, which coincides with the fingerprint ridge period. The high-spatial-resolution distribution of Merkel cells on the fingertips are said to encode fine texture into a temporal code (Weber et al., 2013). To compare with readout areas, we assume that the Merkel cells have an equivalent sensing area equal to $0.4 \times 0.4\,\text{mm}^2$ (based on the equivalent pitch), depicted by the vertical line in Fig. 3.35. The surprisingly large number of prior e-skin work with readout per-taxel area that is smaller than the equivalent sensing area for Merkel cells is a testimony of the massive progress achieved in VLSI technology. The equivalent power consumption

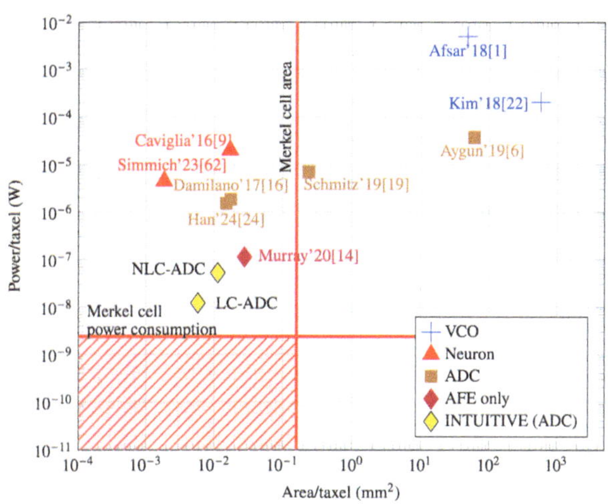

Fig. 3.35 Area per taxel versus power per taxel for e-skin readouts. To match the Merkel cell equivalent performance, texture sensing requires high-spatial-resolution and an ultra-low-power readout. Recent single-unit mapping (Hay & Pruszynski, 2020) shows the subfield pattern of Merkel cells matches the 0.4-mm fingerprint ridge period. The equivalent power consumption is derived based on the required energy to propagate a spike along the axon (see Appendix A)

is derived based on the required energy to propagate a spike along the axon (see Appendix A).

While many of the prior works achieve the area target, these prior works, however, have off-chip sensors. Thus, while the number of readout channels per chip can scale favorably with a smaller per-taxel area, a key bottleneck would be the amount of wiring needed to connect the sensors to the electronics. This limitation has been discussed previously in Sect. 2.2.3. Our work in Alea et al. (2024) and Alea et al. (2023) was the first to integrate sensors on a high-density readout chip, eliminating the wiring limitation, while achieving the required density for high-spatial-resolution texture sensing.

3.3.1.2 Slip Sensing

Si-CMOS designs are constrained by their sizes, often to only a few millimeters. This limitation is due to the fabrication cost and the rigid structure of the chips (a larger area is more prone to damage under bending stress). This is especially suitable for localized texture sensing and discrimination, as performed in the fingertips. Slip sensing, on the other hand, requires large-area coverage to be able to interface with larger macroscale objects. While most of the readout designs implemented on Si-CMOS achieve the stringent requirement for the per-taxel spatial resolution in high-density localized texture sensing. The prior large-area readouts for tactile sensing are off by a factor 100× with the target area of $0.5 \times 0.5\,\text{mm}^2$ (Fig. 3.36). The target area assumes an equivalent pitch of 0.5 mm (Johnson, 2001), as observed for the Meissner corpuscles, which are responsible for shear and strain encoding. Recent microneurography (Delhaye et al., 2021) mapping show that local, ridge-scale (0.5 mm) hotspots encode the shear forces and strain via the Meissner corpuscles. The equivalent power consumption is derived based on the energy required to propagate a spike along the axon (see Appendix A). Palmar regions are estimated to have an equivalent 2-mm pitch (Johansson & Vallbo, 1979).

3.3.1.3 LC-ADCs Are Area-efficient

To achieve true per-taxel sampling with integrated sensors, pitch-matched readout and sensors are crucial. Aside from the inherent asynchronous sampling benefits of LC-ADCs, which bodes well for spike-based readout, the relative simplicity of the LC-ADC architecture (Fig. 3.37) also makes for an area-efficient per-taxel readout. One-bit-DAC LC-ADC implementations, as opposed to multi-bit approaches, have the benefit of a reduced area in exchange for a loop-delay-dependent error (Timmermans et al., 2024). If the comparator is designed to be fast enough, this error is mitigated. Therefore, the area is only set by the two window comparators (see Fig. 3.37). As seen in Figs. 3.35 and 3.36, the area efficiency allows our designs (Alea et al., 2024, 2025) to achieve area constraints set by the desire to meet the biological capabilities (Merkel and Meissner cells). In combination with

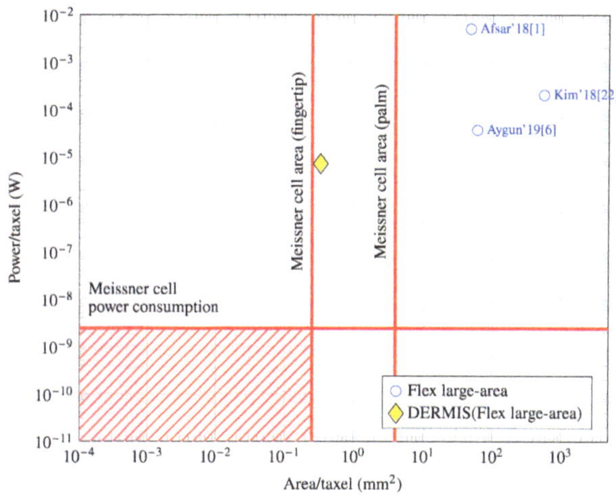

Fig. 3.36 Area per taxel versus power per taxel for e-skin solutions with flexible large-area readout. Slip sensing requires a large-area coverage in both the fingertips and the hand palm regions. Recent microneurography (Delhaye et al., 2021) mapping shows that local, ridge-scale (0.5 mm) hotspots encode the shear strain via Meissner corpuscles. The equivalent power consumption is derived based on the required energy to propagate a spike along the axon (see Appendix A)

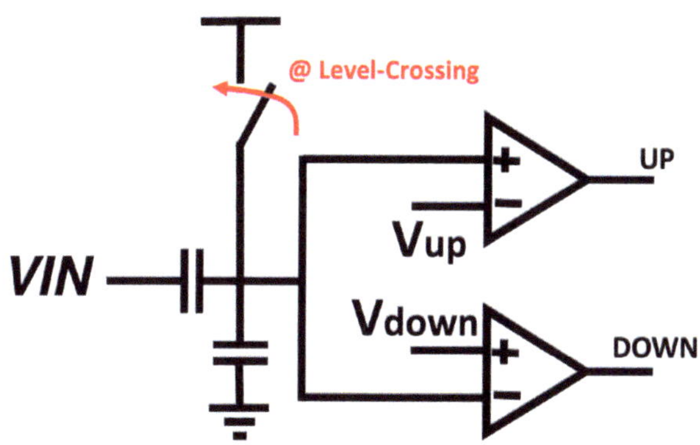

Fig. 3.37 The LC-ADC performs the same event-based sparse sampling mechanism in biological neurons. One-bit DAC LC-ADC implementations, as opposed to multi-bit approaches, have the benefit of a reduced area in exchange of a loop delay-dependent error (Timmermans et al., 2024)

using a finer thin-film technology, the LC-ADC architecture enables more than two orders of magnitude area improvement over previous large-area designs, as seen in Fig. 3.36.

3.3.2 *Power Consumption Challenge*

Biological afferent firing is inherently energy efficient (Ochab-Marcinek et al., 2009). Based on well-known energy per spike estimates for the primary afferents in the human skin, we will compare biology with prior e-skin works. Prior to our work, the state of the art (Murray et al., 2020) in the e-skin readout power consumption is a factor 100× larger than its biological equivalent. This is a significant gap if we want to achieve more human-like performance in next-generation tactile sensor designs. Thus, a key breakthrough is required to reduce this gap, as will be discussed next.

3.3.2.1 SNN-based Processing Enables Low-power Readout

As previously discussed in Sect. 3.1.2, the SNN-based processing of readout outputs allows for the relaxation of the specifications, in terms of noise and mismatch, and overall resolution. Our prior work (Alea et al., 2024) has proven that SNNs are inherently robust to noise and mismatch, resulting in robust classification accuracies despite these errors. Furthermore, we have shown in Alea et al. (2022) that a low spike-encoding resolution of 3–5 bits is not only sufficient, but also avoids overfitting, and thus maximizes the classification accuracy. These observations form the basis for our aggressive power scaling to achieve more human-like levels. This is particularly beneficial for per-taxel readout, since the system power consumption scales linearly with the number of taxels. We show in Fig. 3.35 that our designs reduce the power consumption gap from 100 (Murray et al., 2020) to only around 6. Fundamentally, the power consumption is limited by the required 1 kHz sensing bandwidth to extract the high-frequency texture information from the piezoelectric sensors. Crucially, we note that the work in Murray et al. (2020) only includes the power consumption of the signal conditioning frontend without the ADC. The readout power reported in Fig. 3.35 includes both the piezoelectric sensor charge amplifier and the LC-ADC (Alea et al., 2024).

3.4 Conclusion

In this chapter, we have described our two-pronged approach towards bridging the gap between biological skin and its electronic counterpart: (a) optimized end-to-end design of the e-skin system while (b) leveraging the integration benefits of VLSI technology. Our end-to-end design approach borrows inspiration from biology.

We have detailed our pragmatic technology-aware bioinspired methodology, wherein we have adopted various biological concepts that result in system-level benefits for our e-skin design. Our key takeaways from this approach are as follows: (a) borrowing the temporally-sparse sampling in neuronal firing using spike-based encoding breaks the tradeoff between signal encoding fidelity and power-efficiency, especially during periods of sparse activity, and (b) spike-encoding lends well to *self-correcting* SNN-based classifiers, resulting in relaxed readout design constraints, which fit perfectly the goal of (c) true spatiotemporally-sparse sampling, which now includes both spatial and temporal compression of sparse tactile input.

We have highlighted breakthroughs towards next-generation fully-integrated e-skin. We have shown the successful demonstration of miniaturization and integration of piezoelectric and capacitive shear tactile sensors. We have also highlighted critical design challenges with regards to achieving per-taxel readout, and described a key takeaway that level-crossing-based converters present a power- and area-efficient pathway towards achieving this goal, while maintaining the same time-sparse sampling behavior that we deemed beneficial with our pragmatic bioinspired approach.

3.4.1 Multi-modal and Multi-scale System Approach: The INTUITIVE and DERMIS Designs

To build fully functional and practically useful robotic/prosthetic hands for dexterous manipulation, we will implement a biologically-inspired approach of equipping the hand with a multi-modal and multi-scale e-skin sensing solution. Effectively leveraging the strengths of both high-density Si-CMOS (INTUITIVE) and TFT-based implementations (DERMIS) while addressing the inherent limitations of both approaches, our goal is to demonstrate both localized texture sensing and slip detection with high system-level power efficiency and tactile encoding accuracy. Combining tactile sensors of different modality ((a) piezoelectric sensors for high-spatial-resolution dynamic force sensing for texture scans, and (b) capacitive sensors for static and dynamic shear- and normal-force sensing) follows the multi-modal strategy employed by the human somatosensory system. Our approach to combine both localized small-area texture sensing (via Si-CMOS) with large-area slip sensing (via TFT) employs the same multi-scale approach as in human hands, where fingertips are innervated by high-spatial-resolution mechanoreceptors, while the palmar regions have coarser resolutions but sense over larger areas. In the following chapters, we will discuss the designs and results of our INTUITIVE and DERMIS works.

References

Alea, M. D., Safa, A., Assche, J. V., & Gielen, G. G. E. (2022). Power-efficient and accurate texture sensing using spiking readouts for high-density e-skins. In *2022 IEEE biomedical circuits and systems conference (BioCAS)* (pp. 359–363).

Alea, M. D., Safa, A., Giacomozzi, F., Adami, A., Temel, I. R., Lorenzelli, L., & Gielen, G. (2023). A fingertip-mimicking 12×16 200μm-resolution e-skin taxel readout chip with per-taxel spiking readout and embedded receptive field processing. In *2023 IEEE symposium on VLSI technology and circuits (VLSI technology and circuits)* (pp. 1–2).

Alea, M. D., Safa, A., Giacomozzi, F., Adami, A., Temel, I. R., Rosa, M. A., Lorenzelli, L., & Gielen, G. (2024). A fingertip-mimicking 12×16 200 μm-resolution e-skin taxel readout chip with per-taxel spiking readout and embedded receptive field processing. *IEEE Transactions on Biomedical Circuits and Systems, 18*(6), 1308–1320.

Alea, M. D., Rosa, M. A., Nowicki, Y., Dandekar, M., Myny, K., & Gielen, G. (2025). Dermis: A flexible fully-integrated 600μm-resolution per-taxel slip-to-spikes tactile sensor readout on a-igzo tft for large-area high-density electronic skins. In *2025 Symposium on VLSI Technology and Circuits (VLSI Technology and Circuits)* (pp. 1–3).

Bai, L., Lehnert, B., Liu, J., Neubarth, N., Dickendesher, T., Nwe, P., Cassidy, C., Woodbury, C., & Ginty, D. (2015). Genetic identification of an expansive mechanoreceptor sensitive to skin stroking. *Cell, 163*.

Birznieks, I., Jenmalm, P., Goodwin, A. W., & Johansson, R. S. (2001). Encoding of direction of fingertip forces by human tactile afferents. *Journal of Neuroscience, 21*(20), 8222–8237.

Boutry, C., Negre, M., Jorda, M., Vardoulis, O., Chortos, A., Khatib, O., & Bao, Z. (2018). A hierarchically patterned, bioinspired e-skin able to detect the direction of applied pressure for robotics. *Science Robotics, 3*, eaau6914.

Cheng, M.-Y., Lin, C.-L., & Yang, Y.-J. (2010). Tactile and shear stress sensing array using capacitive mechanisms with floating electrodes. In *2010 IEEE 23rd international conference on micro electro mechanical systems (MEMS)* (pp. 228–231).

Chicca, E., Stefanini, F., Bartolozzi, C., & Indiveri, G. (2014). Neuromorphic electronic circuits for building autonomous cognitive systems. *Proceedings of the IEEE, 102*(9), 1367–1388.

Corniani, G., & Saal, H. (2020). Tactile innervation densities across the whole body. *Journal of Neurophysiology, 124*.

Dahiya, R. S., Cattin, D., Adami, A., Collini, C., Barboni, L., Valle, M., Lorenzelli, L., Oboe, R., Metta, G., & Brunetti, F. (2011). Towards tactile sensing system on chip for robotic applications. *IEEE Sensors Journal, 11*(12), 3216–3226.

Dahiya, R. S., Adami, A., Pinna, L., Collini, C., Valle, M., & Lorenzelli, L. (2014). Tactile sensing chips with posfet array and integrated interface electronics. *IEEE Sensors Journal, 14*(10), 3448–3457.

Damian, D. D., Martinez, H., Dermitzakis, K., Hernandez-Arieta, A., & Pfeifer, R. (2010). Artificial ridged skin for slippage speed detection in prosthetic hand applications. In *Proceedings of the 2010 IEEE/RSJ international conference on intelligent robots and systems (IROS), Taipei, Taiwan, 2010* (pp. 904–909).

Delhaye, B. P., Jarocka, E., Barrea, A., Thonnard, J.-L., Edin, B., & Lefèvre, P. (2021). High-resolution imaging of skin deformation shows that afferents from human fingertips signal slip onset. *eLife, 10*, e64679.

Epoch AI and Our World in Data. (2025). *Computation used to train notable artificial intelligence systems, by domain*. Dataset, Epoch AI, 2025. Major processing by Our World in Data. Original data: "Parameter, Compute and Data Trends in Machine Learning" (Epoch AI). Retrieved February 13, 2026 from https://archive.ourworldindata.org/20260126-094752/grapher/artificial-intelligence-training-computation.html (archived January 26, 2026).

Fernández, R., Payo, I., Vazquez, A., & Becedas, J. (2014). Micro-vibration-based slip detection in tactile force sensors. *Sensors, 14*(1), 709–730.

Frenkel, C., Lefebvre, M., Legat, J.-D., & Bol, D. (2019). A 0.086-mm^2 12.7-pj/sop 64k-synapse 256-neuron online-learning digital spiking neuromorphic processor in 28-nm CMOS. *IEEE Transactions on Biomedical Circuits and Systems, 13*(1), 145–158.

Gerstner, W., Kistler, W. M., Naud, R., & Paninski, L. (2014). *Neuronal dynamics: From single neurons to networks and models of cognition*. Cambridge University Press.

Halata, Z. (1975). The mechanoreceptors of the mammalian skin: ultrastructure and morphological classification. *Advances in Anatomy, Embryology and Cell Biology, 50*(3), 1–77. Comprehensive ultrastructural study of Merkel cells and touch domes in mammalian skin.

Hay, E., & Pruszynski, A. (2020). Orientation processing by synaptic integration across first-order tactile neurons. *PLOS Computational Biology, 16*, e1008303.

He, Y., et al. (2021). A 28.2 μc neuromorphic sensing system featuring SNN-based near-sensor computation and event-driven body-channel communication for insertable cardiac monitoring. In *2021 IEEE Asian solid-state circuits conference (A-SSCC)* (pp. 1–3). IEEE.

Hebb, D. O. (1949). *The organization of behavior: A neuropsychological theory*. Wiley.

Hodgkin, A. L., & Huxley, A. F. (1952). A quantitative description of membrane current and its application to conduction and excitation in nerve. *The Journal of Physiology, 117*(4), 500–544.

Hohendorff, B. (2019). *Jamming of fingers: An experimental study to determine forces and fracture thresholds*. Ph.D Thesis. Reports average maximum force of 42 N applied to the index finger in human participants.
Indiveri, G. (2025). Neuromorphic is dead. long live neuromorphic. *Neuron, 113*(20), 3311–3314. NeuroView.
Janotte, E., Bamford, S., Richter, O., Valle, M., & Bartolozzi, C. (2022). Neuromorphic capacitive tactile sensors inspired by slowly adaptive mechanoreceptors. In *2022 20th IEEE interregional NEWCAS conference (NEWCAS)* (pp. 119–123).
Johansson, R., & Flanagan, J. (2009). Coding and use of tactile signals from the fingertips in object manipulation tasks. *Nature Reviews. Neuroscience, 10*, 345–359.
Johansson, R., & Vallbo, A. (1979). Tactile sensibility in the human hand: Relative and absolute density of four types of mechanoreceptive units in glabrous skin. *The Journal of Physiology, 286*, 283–300.
Johnson, K. O. (2001). The roles and functions of cutaneous mechanoreceptors. *Current Opinion in Neurobiology, 11*(4), 455–461.
Johnson, P. W., & Blackstone, J. M. (2007). Children and gender—differences in exposure and how anthropometric differences can be incorporated into the design of computer input devices. *Scandinavian Journal of Work, Environment & Health, Supplement, 3*, 26–32. Includes index finger width values used in device design; Table 1 reports adult index finger width (mean and SD).
Jörntell, H., Bengtsson, F., Geborek, P., Spanne, A., Terekhov, A., & Hayward, V. (2014). Segregation of tactile input features in neurons of the cuneate nucleus. *Neuron, 83*.
Kang, X., Yang, Z., Ren, Y., & Ye, T. T. (2025). Spike-count reduction techniques for low power spiking neural networks. *Neural Processing Letters, 57*, 75.
Khamis, H., Redmond, S., Macefield, V., & Birznieks, I. (2014). Tactile afferents encode grip safety before slip for different frictions. In *Proceedings of the 36th annual international conference of the IEEE engineering in medicine and biology society (EMBC 2014): Discovering, innovating, and engineering future biomedicine, 26-30 August 2014, Chicago, Illinois* (pp. 4123–4126). IEEE. IEEE Engineering in Medicine and Biology Society. Annual International Conference ; Conference date: 11-07-2022.
Khamis, H., Izquierdo Albero, R., Salerno, M., Shah Idil, A., Loizou, A., & Redmond, S. J. (2018). Papillarray: An incipient slip sensor for dexterous robotic or prosthetic manipulation—design and prototype validation. *Sensors and Actuators A: Physical, 270*, 195–204.
Kingma, D. P., & Ba, J. (2014). *Adam: A method for stochastic optimization*. Preprint. arXiv:1412.6980
Kylberg, G. (2011). The kylberg texture dataset v. 1.0. Technical Report 35, Centre for Image Analysis, Swedish University of Agricultural Sciences and Uppsala University, Uppsala, Sept. 2011. Available online at http://www.cb.uu.se/~gustaf/texture/ and https://kylberg.org/datasets/.
Laughlin, S. B. (2001). Energy as a constraint on the coding and processing of sensory information. *Current Opinion in Neurobiology, 11*(4), 475–480.
Lee, W. W., Kukreja, S. L., & Thakor, N. V. (2017). Discrimination of dynamic tactile contact by temporally precise event sensing in spiking neuromorphic networks. *Frontiers in Neuroscience, 11-2017*.
Lesniak, D. R., Marshall, K. L., Wellnitz, S. A., Jenkins, B. A., Baba, Y., Rasband, M. N., Gerling, G. J., & Lumpkin, E. A. (2014). Computation identifies structural features that govern neuronal firing properties in slowly adapting touch receptors. *eLife, 3*, e01488.
Lichtsteiner, P., Posch, C., & Delbruck, T. (2008). A 128$\times$ 128 120 db 15 μs latency asynchronous temporal contrast vision sensor. *IEEE Journal of Solid-State Circuits, 43*(2), 566–576.
Marisa, T., Niederhauser, T., Haeberlin, A., Wildhaber, R. A., Vogel, R., Goette, J., & Jacomet, M. (2017). Pseudo asynchronous level crossing adc for ecg signal acquisition. *IEEE Transactions on Biomedical Circuits and Systems, 11*(2), 267–278.
McCulloch, W. S., & Pitts, W. H. (1943). A logical calculus of the ideas immanent in nervous activity. *The Bulletin of Mathematical Biophysics, 5*(4), 115–133.

Mensi, S., Naud, R., Pozzorini, C., Avermann, M., Petersen, C. C., & Gerstner, W. (2012). Parameter extraction and classification of three cortical neuron types reveals two distinct adaptation mechanisms. *Journal of Neurophysiology, 107*(6), 1756–1775.

Murray, S. J., Medinger, J., Schmitz, J., Balkır, S., & Hoffman, M. W. (2020). Toward a low power e-skin interface system on a chip for taxel arrays. In *2020 IEEE International Symposium on Circuits and Systems (ISCAS)* (pp. 1–5).

Nierula, B., Stephani, T., Bailey, E., Kaptan, M., Pohle, L.-M. G., Horn, U., Mouraux, A., Maess, B., Villringer, A., Curio, G., Nikulin, V. V., & Eippert, F. (2024). A multichannel electrophysiological approach to noninvasively and precisely record human spinal cord activity. *PLOS Biology, 22*(10), 1–41.

Nyquist, H. (1928). Certain topics in telegraph transmission theory. *Transactions of the American Institute of Electrical Engineers, 47*(2), 617–644.

Ochab-Marcinek, A., Schmid, G., Goychuk, I., & Hänggi, P. (2009). Noise-assisted spike propagation in myelinated neurons. *Physical Review E, 79*, 011904.

Pfeiffer, M., & Pfeil, T. (2018). Deep learning with spiking neurons: Opportunities and challenges. *Frontiers in Neuroscience, 12*, 774.

Purves, D., Augustine, G. J., Fitzpatrick, D., Katz, L. C., LaMantia, A.-S., McNamara, J. O., & Williams, S. M. (2001). The refractory period. In *Neuroscience* (2nd ed.). Sinauer Associates. NCBI Bookshelf ID: NBK11146, Chapter in: Voltage-Dependent Membrane Permeability.

Romeo, R. A., Oddo, C. M., Carrozza, M. C., Guglielmelli, E., & Zollo, L. (2017). Slippage detection with piezoresistive tactile sensors. *Sensors, 17*(8).

Rosa, M. B. A., & Kraft, M. (2024). (2024). Microengineered flexible pressure sensors with sacrificial molding layer: A novel fabrication approach for improved performance. *Proceedings, 97*(1).

Rosenblatt, F. (1958). The perceptron: A probabilistic model for information storage and organization in the brain. *Psychological Review, 65*(6), 386–408.

Safa, A., Corradi, F., Keuninckx, L., Ocket, I., Bourdoux, A., Catthoor, F., & Gielen, G. G. E. (2023). Improving the accuracy of spiking neural networks for radar gesture recognition through preprocessing. *IEEE Transactions on Neural Networks and Learning Systems, 34*(6), 2869–2881.

Shannon, C. E. (1949). Communication in the presence of noise. *Proceedings of the IRE, 37*(1), 10–21.

Su, Z., Hausman, K., Chebotar, Y., Molchanov, A., Loeb, G. E., Sukhatme, G. S., & Schaal, S. (2015). Force estimation and slip detection/classification for grip control using a biomimetic tactile sensor. In *2015 IEEE-RAS 15th international conference on humanoid robots (humanoids)* (pp. 297–303).

Timmermans, M., van Oosterhout, K., Fattori, M., Harpe, P., Liu, Y.-H., & Cantatore, E. (2024). A 1.8–65 fj/conv.-step 64-db sndr continuous- time level crossing adc exploiting dynamic self-biasing comparators. *IEEE Journal of Solid-State Circuits, 59*(4), 1194–1203.

Tremblay, M. R., & Cutkosky, M. R. (1993). Estimating friction using incipient slip sensing during a manipulation task. In *Proceedings of the 1993 IEEE international conference on robotics and automation (ICRA), Atlanta, GA, USA, May 1993* (Vol. 1, pp. 429–434).

Tsividis, Y. (2010). Event-driven data acquisition and continuous-time digital signal processing. In *IEEE Custom Integrated Circuits Conference 2010* (pp. 1–8).

Ulloa, P. M., Bulens, D. C., & Redmond, S. J. (2022). Incipient slip detection for rectilinear movements using the papillarray tactile sensor. In *2022 IEEE Sensors* (pp. 1–4).

Van Assche, J., & Gielen, G. (2020). Power efficiency comparison of event-driven and fixed-rate signal conversion and compression for biomedical applications. *IEEE Transactions on Biomedical Circuits and Systems, 14*(4), 746–756.

Van Assche, J., & Gielen, G. (2024). Analysis and design of a 10.4-enob 0.92–5.38-μw event-driven level-crossing adc with adaptive clocking for time-sparse edge applications. *IEEE Journal of Solid-State Circuits, 59*(9), 2858–2869.

Van Assche, J., Frenkel, C., Safa, A., & Gielen, G. (2025). Freya: A 0.023-mm^2/channel, 20.8- μw/channel, event-driven 8-channel soc for spiking end-to-end sensing of time-sparse biosignals. *IEEE Transactions on Circuits and Systems I: Regular Papers, 72*(3), 1093–1104.

Wang, T., Wang, D., Hurst, P. J., Levy, B. C., & Lewis, S. H. (2009). A level-crossing analog-to-digital converter with triangular dither. *IEEE Transactions on Circuits and Systems I: Regular Papers, 56*(9), 2089–2099.

Wang, Y., Xi, K., & Mei, D. (2016). Slip detection in prosthetic hand grasping by using the discrete wavelet transform analysis. In *Proceedings of the 2016 IEEE international conference on advanced intelligent mechatronics (AIM), Banff, AB, Canada, 2016* (pp. 1485–1490).

Ward-Foxton, S. (2020). Le cun: "it's really hard to succeed with exotic hardware". *EE Times Europe*. Interview/coverage quoting Yann Le Cun: "There is a need to prove that the algorithms work before building chips for them.".

Weber, A. I., Saal, H. P., Lieber, J. D., Cheng, J.-W., Manfredi, L. R., III, J. F. D., & Bensmaia, S. J. (2013). Spatial and temporal codes mediate the tactile perception of natural textures. *Proceedings of the National Academy of Sciences of the United States of America, 110*(42), 17107–17112.

Weltin-Wu, C., & Tsividis, Y. (2013). An event-driven clockless level-crossing adc with signal-dependent adaptive resolution. *IEEE Journal of Solid-State Circuits, 48*(9), 2180–2190.

Yoon, Y. C. (2017). Lif and simplified srm neurons encode signals into spikes via a form of asynchronous pulse sigma–delta modulation. *IEEE Transactions on Neural Networks and Learning Systems, 28*(5), 1192–1205.

Yousuf, M., Arya, D. S., Garg, M., Varshney, D., & Singh, P. (2022). Asymmetric capacitive pixel array (e-cap) for mapping of epidermal vectored movement. *IEEE Journal on Flexible Electronics, 1*(2), 134–140.

Yuan, W., Li, R., Srinivasan, M. A., & Adelson, E. H. (2015). Measurement of shear and slip with a gelsight tactile sensor. In *2015 IEEE International Conference on Robotics and Automation (ICRA)* (pp. 304–311).

Zambrano, D., & Bohte, S. M. (2016). Fast and efficient asynchronous neural computation with adapting spiking neural networks. Preprint, arXiv:1609.02053. https://doi.org/10.48550/arXiv.1609.02053

Zambrano, D., et al. (2019). Sparse computation in adaptive spiking neural networks. *Frontiers in Neuroscience, 13*, 987.

Zangrandi, A., Picard, A., Terekhov, A. V., Thonnard, J.-L., Lefèvre, P., & Hayward, V. (2021). Neurophysiology of slip sensation and grip reaction: insights for hand prosthesis control of slippage. *Journal of Neurophysiology, 126*(2), 477–492.

Chapter 4
INTUITIVE: A Fingertip-Mimicking 12 x 16 200 μm-Resolution *E*-skin Taxel Readout Chip with Per-Taxel Spiking Readout and Embedded Receptive Field Processing

Abstract This chapter presents the first on-chip, per-taxel sensor-to-spike *e*-skin system with embedded complex receptive field (CRF) processing at fingertip-inspired taxel density, achieving an unprecedented spatial resolution of 200 μm. This chapter introduces the INTUITIVE chip, integrating a miniaturized 12×16 polyvinylidene fluoride (PVDF) piezoelectric sensor array on a silicon CMOS readout chip with dual per-taxel spiking readout channels. We demonstrate that SNN-based tactile classification is relatively *insensitive* to circuit non-idealities, enabling relaxed circuit targets that reduce area and power while still meeting strict per-taxel readout requirements. Experimental results show excellent spatiotemporal spiking classification accuracies for texture, flutter, and indentation frequency using only a 256-neuron SNN classifier, a 3–5 bit equivalent spike-encoding resolution, and a sub-Nyquist population spiking rate, and further indicate that the proposed neuromorphic level-crossing sampling (N-LCS) with exponentially decaying threshold outperforms conventional bipolar LCS with fixed threshold. Fabricated in 0.18-μm CMOS, the prototype achieves 12.33 nW per taxel (75 μW to 5 mW total, input-dependent) and enables near–human-level texture sensing by surpassing Merkel-cell spatial resolution (0.2 versus 0.4 mm) while improving power efficiency by more than 10×.

Keywords Level-crossing ADC · Per-taxel readout · Receptive field processing · Spike-based processing

In the previous chapters, the end-to-end integration towards achieving more human-like tactile sensing for future next-generation tactile systems. The previous chapter has also presented the advantages of various bioinspired techniques towards next-generation e-skin systems: (a) spike-based encoding, (b) SNN-based processing, (c) local processing. We have demonstrated that neuron-inspired spike-based processing is not only a power-efficient way to encode the tactile stimuli, but also improves on prior event-based signal acquisition techniques. We have also demonstrated that the inherent robustness of SNN-based classification accuracies to readout non-

M. D. Alea, G. Gielen, *Chips That Feel*, Analog Circuits and Signal Processing,
https://doi.org/10.1007/978-3-032-26584-5_4

idealities helps to relax the readout specifications in order to achieve a lower power consumption and a smaller chip area.

This chapter will demonstrate in the INTUITIVE chip, for the first time, the above hypothesis. It demonstrates a first-of-its-kind fully-integrated e-skin readout chip, integrating a state-of-the-art 200-μm spatial resolution with a dense 12×16 piezoelectric sensor array. This work mimics the fine sensing and first-order processing seen in the human fingertips. In contrast to previous *e*-skin implementations where low-spatial resolution sensors are off-chip and the sensor data are acquired per column and digitized in a frame-based manner, this INTUITIVE chip is the first demonstration of a high-density *e*-skin readout chip with:

- per-taxel asynchronous *spiking* readout,
- mimicked on-chip complex receptive field *(CRF) processing*, and
- *integrated* piezoelectric sensors, while
- achieving a *fingertip-mimicking* state-of-the-art taxel spatial resolution of 0.2-mm.

4.1 INTUITIVE Chip: Motivation

To mimic the rich tactile information captured by the human fingertips during object manipulation, it is necessary to achieve the same power-efficient and high-spatial-resolution sensing and local processing. To achieve this, we have proposed to integrate a dense tactile sensor array on a custom CMOS readout chip, allowing for a dense and power-efficient readout with local processing capabilities. The sensor-to-readout chip integration supports dense per-taxel connections, enabling *true* per-taxel sparse-sampling conversion (i.e., spikes) as performed in human afferents (Fig. 4.1c). Prior frame-based readout approaches, discussed in Chap. 2, are illustrated in Fig. 4.1a and b. For such sensor integration, polyvinylidene fluoride (PVDF)-based piezoelectric sensors, as discussed in Chap. 3, are a good solution due to their relatively simple structure and their CMOS-compatible fine lithography-based integration (Schmitz et al., 2019; Murray et al., 2020; Pinna et al., 2017; Dahiya et al., 2014). Piezoelectric sensors are sensitive to high-frequency force, therefore are catered to applications where acquiring the magnitude of the static force is of lesser importance, such as in *texture* or *vibration* sensing. Discerning between objects through the delicate sensory perception of fine textures on human fingertips commonly involves scanning textured objects, inducing vibrations that are subsequently translated into spatio-temporal spike patterns by mechanoreceptors on the skin (Weber et al., 2013).

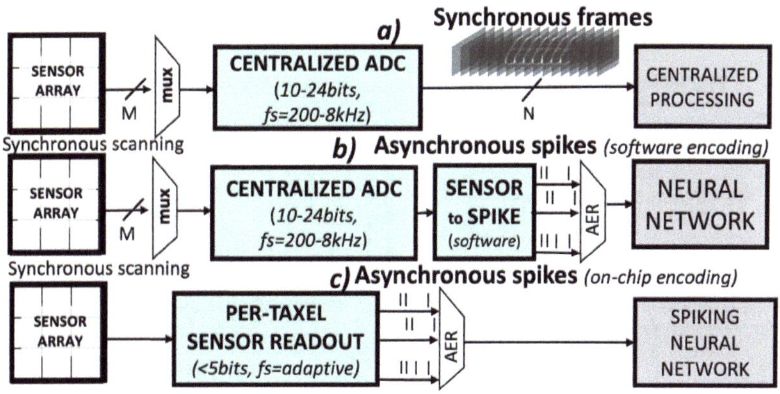

Fig. 4.1 Existing state-of-the-art electronic skin readouts can be classified into two major categories: (**a**) synchronous frame generation using a central ADC and further digital processing; (**b**) asynchronous spike generation in software followed by a spiking neural network. (**c**) The proposed novel *e*-skin design with on-chip per-taxel sensor-to-spike conversion

4.2 Piezoelectric Sensor Integration

The PVDF-based sensors are integrated on standard commercial CMOS wafers, while achieving the smallest sensor pitch ever (200 μm) and the thinnest dielectric layer (250 nm). These innovations have enabled (a) an increase in the sensing spatial resolution, (b) a reduction in the required high voltage applied during poling (25 V for a 100 V/μm layer thickness), and (c) the use of a higher-density 0.18 μm CMOS process fabricated in commercial technologies for readout, digital processing, and communication.

The sensor structure (Fig. 4.2a) is realized by spin coating a PVDF solution of PVDF-TrFE (Piezotech FC20) on quadrants of the 8" 0.18-μm CMOS wafer. The patterned top electrode (*Au*, thickness = 150 nm) also serves as a mask for the PVDF lithography. The chip's top metal serves as the bottom electrode.

4.3 Chip and Taxel Architecture

The taxel array pitch of $200 \times 200\,\mu m^2$ is set by the resolution of the sensor's wafer-based post-processing lithography. The implemented per-taxel readout circuits occupy smaller areas, enabling the INTUITIVE chip to integrate both a standard level-crossing converter and a neuromorphic level converter within a single taxel. In our prototype chip, for comparison reasons, both a voltage-mode level-crossing sampling (LCS), and a *neuromorphic* current-mode LCS (N-LCS) converter readout have been implemented. To eliminate sensor-related discrepancies between the performances of the two spiking-based readouts, both architectures are integrated

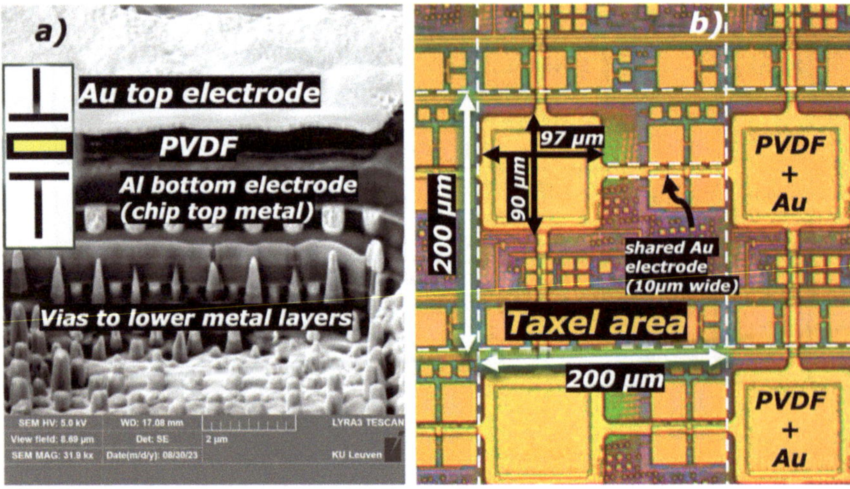

Fig. 4.2 Integrated PVDF tactile sensors: (**a**) SEM image showing the cross-section of a single post-processed tactile sensor for a PVDF thickness of 250 nm with the cross-section diagram of the sensor structure superimposed, (**b**) Chip micrograph of the tactile sensors post-processed on top of the CMOS chip, showing the taxel pitch of 200 µm

under the same taxel, thereby sharing the same set of sensors. The dual per-taxel readout design also allows for flexibility in the spike generation depending on the application. For instance, in robotics applications, a simple LCS sensor-to-spike conversion might be sufficient for upstream processing, while for *neuroprosthetics*, the more biologically faithful spike trains from the N-LCS channel might work better. The LCS accepts a *voltage* input (charge amplifier output) while the N-LCS accepts a *current* input coming from the output of a per-taxel voltage-to-current (V-to-I) converter.

Of the 12×16 N-LCS channels in the prototype chip, 24 are special bio-mimicking spiking units called *heminodes* (Fig. 4.3b and c), which sum multiple currents from different taxels before generating spikes, forming user-configurable CRFs. Each heminode input is connected to each of the 192 V-to-I outputs through an analog demultiplexer switch matrix, allowing for programmable sensor-to-heminode connections. The per-taxel demultiplexer configuration is then stored in digital user-configurable on-chip registers. The summation of sensor values mimics the formation of human fingertip CRFs (Fig. 4.3a). The chip's overall architecture is shown in Fig. 4.4. For chip testing, an on-chip replica frontend chain is used. The per-taxel frontend amplifiers and the level-crossing comparators, operating in weak inversion, are biased with a PTAT current to compensate for the circuit's temperature dependence.

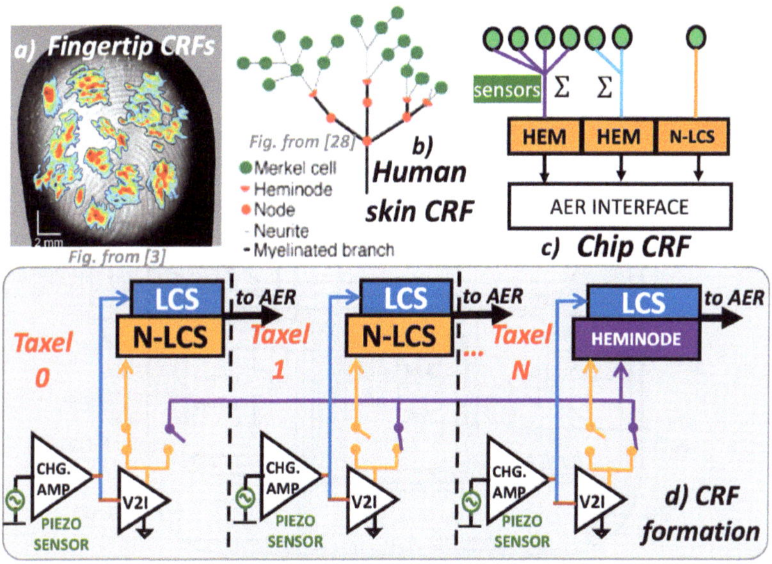

Fig. 4.3 System architecture. (**a**) Human fingertip CRFs are formed by (**b**) combining Merkel cell outputs to spike-inducing biological heminodes (Lesniak et al., 2014), and is (**c**) mimicked on chip by summing sensor outputs into neuromorphic heminodes. (**d**) The chip consists of dual channel spiking readout per taxel. The LCS accepts the charge amplifier output (*voltage*) while the N-LCS/heminodes accept the V-to-I output (*current*). The per-taxel V-to-I output can be individually routed on either: *(i)* the local N-LCS taxel or, *(ii)* any of the 24 heminode lines (ex: *purple*) through the configurable switches. The summation of sensor values mimic the Merkel cell-heminode junctions seen in human heminodes. The summed current befores it enters the spike generation unit of the heminode is regularized/normalized via a current mirror with digitally-programmable gain ratio

4.4 Spatial Compression via Heminodes

We modeled the effects of various complex receptive field (CRF) shapes, enabled by on-chip heminodes, to the texture classification accuracy. Classification is performed using the Kylberg texture dataset (Kylberg, 2011), using the sensor-to-spike framework used in Chap. 3. Heminodes branching out to both global and local portions of the entire 196-taxel emulated array are modeled. These heminodes are randomly-innervated with: (a) a uniformly-distributed number of taxels per heminode (i.e., equal taxels per heminode), or (b) a skewed distribution (i.e., heminodes are innervated with a progressively-decreasing number of taxels, as observed in biological studies in Lesniak et al. 2014). The configurations are summarized and depicted in Fig. 4.5.

The results of the modeling include heminode counts of both 32 and 128, as summarized in Fig. 4.6. The key takeaways are: (a) that increasing the heminode count from 32 to 128 generally increases the classification accuracy, (b) heminodes with access to the entire array (global) and with skewed distribution achieve

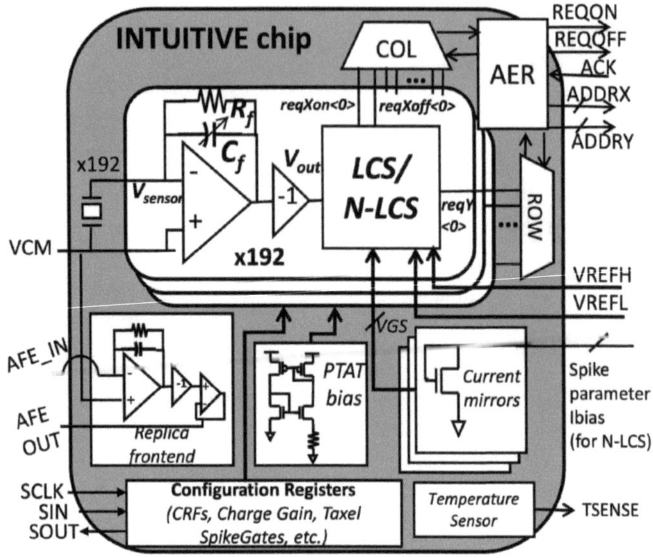

Fig. 4.4 The INTUITIVE e-skin taxel readout chip architecture. The chip has a 12×16 taxel array

the highest classification accuracies, and (c) despite the low heminode count, the heminode with global and skewed distribution innervations achieves excellent classification accuracies, which are only matched by the alternative uniformly-distributed innervations at the higher heminode count of 128. Keeping the heminode count low achieves higher spatial compression.

4.5 Circuit Implementation

4.5.1 Design Strategy and Error Sources

The per-taxel nature of the spiking readout circuits would impose tight targets on the area and power consumption. Fortunately, as shown in Sect. 3.1.2, the *SNN-based* classification performance is relatively *insensitive* to the circuit non-idealities (Fig. 3.16). This allows to *relax* the circuit error targets, as enumerated below, in favor of a reduced area and power consumption.

In this design, the circuit error sources are divided into two categories: a) *time-invariant* (random and systematic) and b) *time-variant.* Random time-invariant errors mostly come from device mismatches resulting in inter-taxel mismatches on the charge gain, the high- and low-pass corner frequencies, the spiking thresholds, and DC-level signal offset (due to the random input-referred offset). Sensor mismatch (due to the fabrication) can also be grouped together with the other random

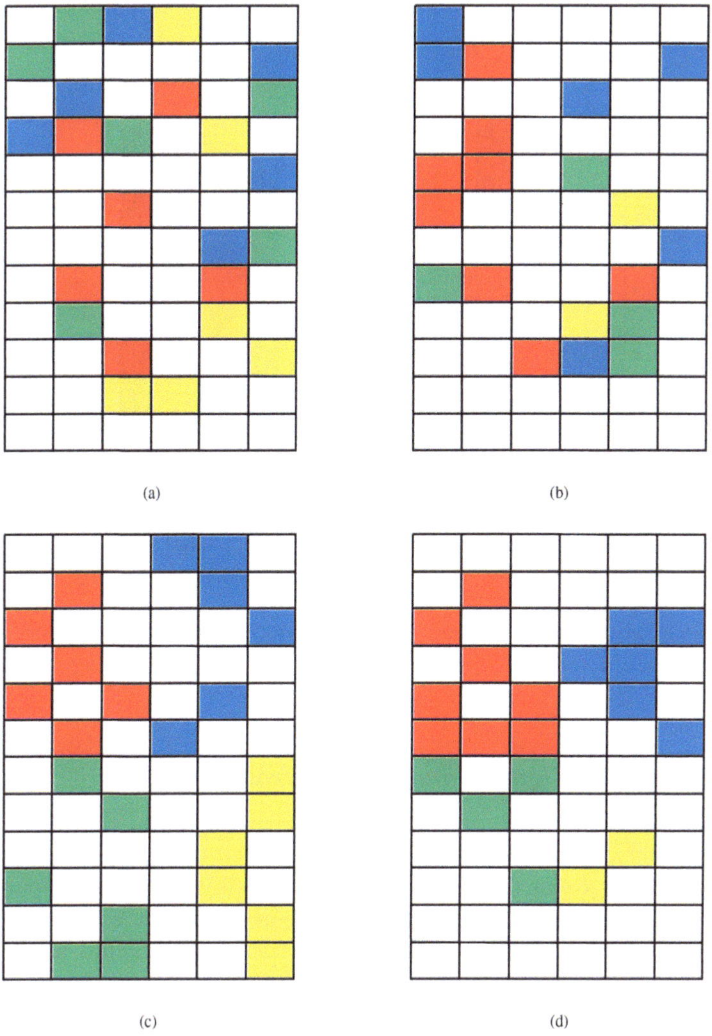

Fig. 4.5 CRF configurations: (a) global & uniform distribution, (b) global & skewed distribution, (c) local & uniform distribution, (d) local & skewed distribution. Each color (four heminodes) represents taxel-to-heminode *innervations* or connections of the same heminode

time-invariant errors. Systematic time-invariant errors, such as static and dynamic settling errors, are mainly due to the finite OTA open-loop gain, gain-bandwidth, and systematic offset. Random and systematic time-invariant errors, if assumed to have only a small drift, can inherently be *calibrated* out by training the classifier used on the readout output with the same fixed errors. The classification accuracy's insensitivity to random time-variant error sources, such as the device thermal and

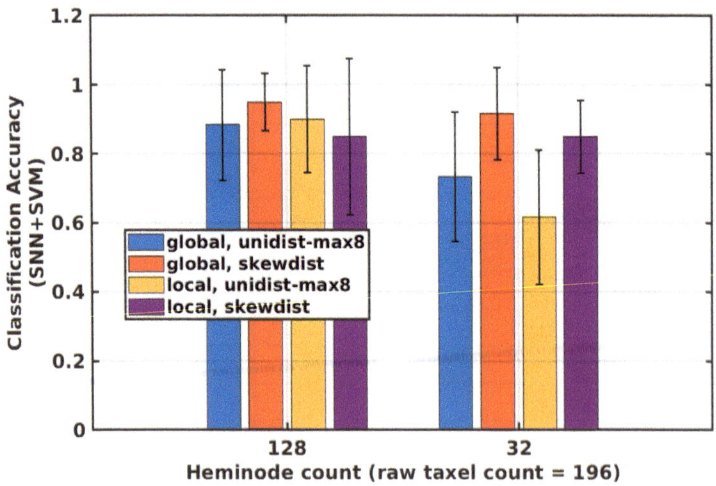

Fig. 4.6 Modeling the effect of heminode innervation topology (to form CRFs) to the texture classification accuracy. The emulated sensors are part of a 14×14 taxel array

flicker noise, also allowed for a design with a reduced power consumption and area (see Chap. 3).

4.5.2 Per-taxel Signal Conditioning Frontend

The conditioning frontend chain (Fig. 4.7) consists of: (a) a charge amplifier with configurable charge gain (1.67–25 V/pC), (b) a capacitively-coupled amplifier with an inverting unity gain to cancel the first-stage negative gain, and (c) a low-area voltage-to-current conversion for the N-LCS channel. The circuits are described below.

To reduce the area and power consumption, the first two stages use a simple differential-pair OTA operating in weak inversion (configurable $I_{TAIL} = 1$ nA to 50 nA). The OTAs are biased with a PTAT current to cancel the temperature dependence ($Gm \propto I_{TAIL}/V_T$, where $V_T = kT/q$). The single-stage design and the low bias current of the OTA result in a reduced open-loop gain ($A_{v,openloop} = 140\ V/V$) and gain-bandwidth (15.5 kHz at $I_{TAIL} = 1$ nA), respectively. The small feedback capacitors allow for a large charge gain and a larger GBW, at the expense of increased variability. The input-pair transistors have been sized moderately (see Fig. 4.7) to reduce their area and input parasitic capacitance, which introduces capacitive loading with the sensor charge input, at the expense of increased offset and flicker noise. For a target closed-loop gain of ~ 4, a static gain error of 4.6% is computed, while for an equivalent sampling requirement of $fs = 2^N \cdot \pi \cdot BW_{input}$ (Marisa et al., 2017), where $N = N_{window} = 4\ bits$ and $BW_{input} = 1$ kHz, a 3.5-bit

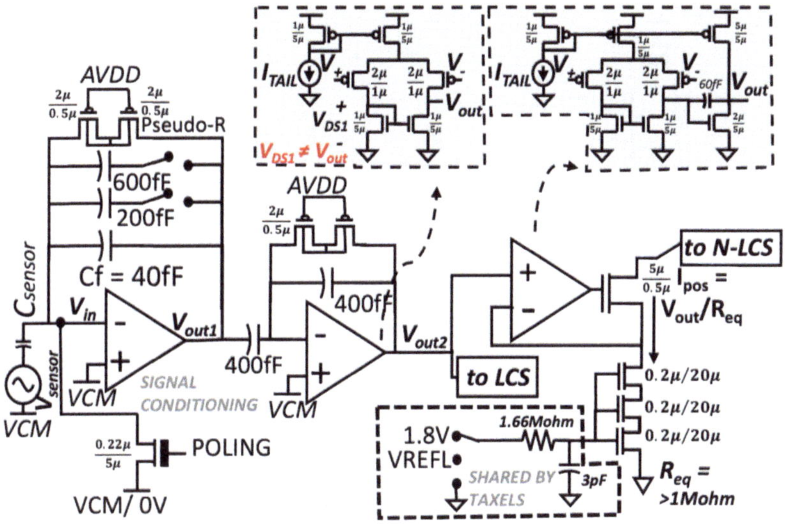

Fig. 4.7 Circuit schematic of the per-taxel signal conditioning readout frontend

resolution is achieved. The effect of the systematic offset due to the OTA current-mirror V_{DS} mismatch is reduced due to the DC-insensitive level-crossing nature of the spike generation (Fig. 4.7).

Despite using piezoelectric tactile sensors, which traditionally are used for detecting dynamic forces, it is also possible to detect low-frequency force components by lowering the charge-amplifier high-pass corner (f_{HPF}). Using back-to-back PMOS-based pseudoresistors (Guglielmi et al., 2020), $f_{HPF} < 0.1\ Hz$ is achieved. The leakage current through the *NWELL-PSUB* junction of the pseudoresistor introduces up to 200 mV of DC-level shift at the output of the amplifiers, but only shifts the overall frontend gain by $< 2.6\%$. The mentioned leakage current has slight photosensitivity and temperature sensitivity, hence it is not constant and changes very slowly. Since the level-crossing conversion "filters out" slow inputs (Fig. 4.8), the effect on the sensor-to-spike conversion is minimized.

The low-pass corner is set by the OTA unity-gain frequency: $f_{LPF} = \beta \cdot \omega_{unity}$. To allow for passband frequency range tunability, I_{TAIL} can be increased at the expense of an increased power consumption ($\omega_{unity} \propto I_{TAIL}/C_L$).

The per-taxel voltage-to-current converter is designed to enable low-current (nA) operation for the N-LCS taxel readout. To achieve an equivalent $R_{V2I} > 1\ M\Omega$ on the restricted area budget while achieving an acceptable V-to-I conversion linearity, the equivalent drain-source impedance of transistors in the triode region is used. The R_{V2I} value can be increased to around 1.6 $G\Omega$ (Fig. 4.9a) to enable ultra-low-current operation (subthreshold region) by decreasing the transistor gate voltage bias $VREFL$ to 0.25 V (Fig. 4.7). This allows reducing the required area for the resistor by around 1500×, as opposed to using high-resistivity poly-resistors. The gate bias is tied to 0 V during $V_{MODE} = \text{'1'}$, where the N-LCS channel is disabled,

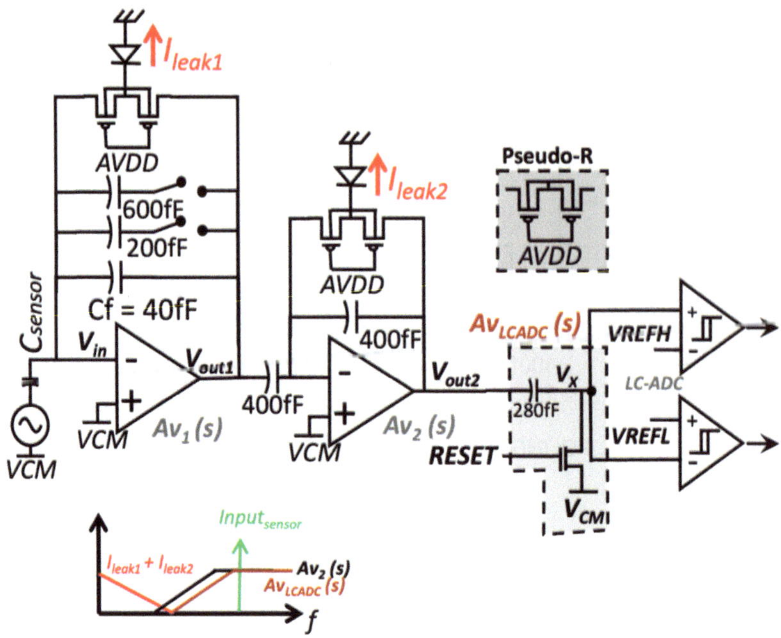

Fig. 4.8 The leakage current through the *NWELL-PSUB* junction of the pseudoresistor introduces up to 200 mV of DC-level shift at the output of the amplifiers, but only shifts the overall frontend gain by < 2.6%. This leakage current has slight photosensitivity and temperature sensitivity, hence it is not constant and changes very slowly. Since the level-crossing conversion "filters out" slow inputs, the effect on the sensor-to-spike conversion is minimal

to minimize its contribution to the per-taxel power consumption. The equivalent resistance R_{V2I}, simulated with 200 Monte-Carlo samples, for $V_{OUT2} = 250\,\text{mV}$ (Fig. 4.7) is shown for $VGATE = \{1.8\,\text{V}, 1\,\text{V}, 0.25\,\text{V}\}$ in Fig. 4.9a, while the simulated linearity of the current output is shown in Fig. 4.9b. These demonstrate an acceptable R_{V2I} spread and linearity despite not using on-chip resistors.

4.5.3 The Voltage-mode LC Channel

The voltage-mode LC channel detects input signal changes and generates bipolar spikes through two window comparators (Fig. 4.10a). An ON spike is generated when $V_{SIG} > V_{REFH}$, while an OFF spike is generated when $V_{SIG} < V_{REFL}$. The window comparator (differential pair with inverter gain stage) allows a scalable biasing for speed-power trade-off ($I_{TAIL} = 1\,\text{nA} - 50\,\text{nA}$). The signal V_{LCS} is reset to V_{CM} after a level crossing. The comparator offset spread, simulated through a Monte-Carlo simulation with 200 samples (Fig. 4.10b), is still sufficiently low, even for a level-crossing window of 62.5 mV ($N = 5\ bits$), the window used in this

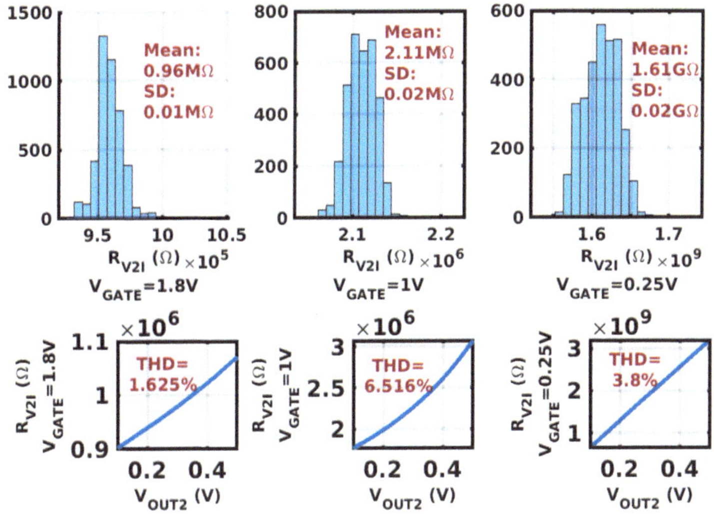

Fig. 4.9 Per-taxel voltage-to-current conversion: (**a**) simulated R_{V2I} at three different gate bias, (**b**) simulated R_{V2I} vs V_{OUT2} (Fig. 4.7) and simulated THD for V_{OUT2} = 10 Hz, 140 mV$_{pp}$ (Q_{IN}= 5 fC$_{pp}$ for C_{sensor}= 172 fF)

work. Considering the comparator mismatch and the AFE noise (which dominates the error sources during the level-crossing conversion), the window size can be set to as low as 12.45 mV (6.3 bits resolution):

$$\sigma_{\text{error}} = \sqrt{e^2_{\text{noise,AFE}} + \sigma^2_{\text{mismatch}}} = \sqrt{(2.97\,\text{mV})^2 + (2.9\,\text{mV})^2} = 4.15\,\text{mV} \tag{4.1}$$

$$Windowsize_{\text{minimum}} = 3 \cdot \sigma_{\text{error}} = 12.45\,\text{mV} \tag{4.2}$$

4.5.4 The Current-mode Neuromorphic LC Channel

The current-mode N-LCS channel outputs unipolar spikes (i.e., ON spikes only) through a delta modulator (Fig. 4.11a). The N-LCS's output spikes are integrated through two current-mode low-pass filters (Chicca et al., 2014) (with fast and slow time constants), forming an exponentially decaying threshold (I_X) for spiking and mimicking biological spike encoding (Mensi et al., 2012).

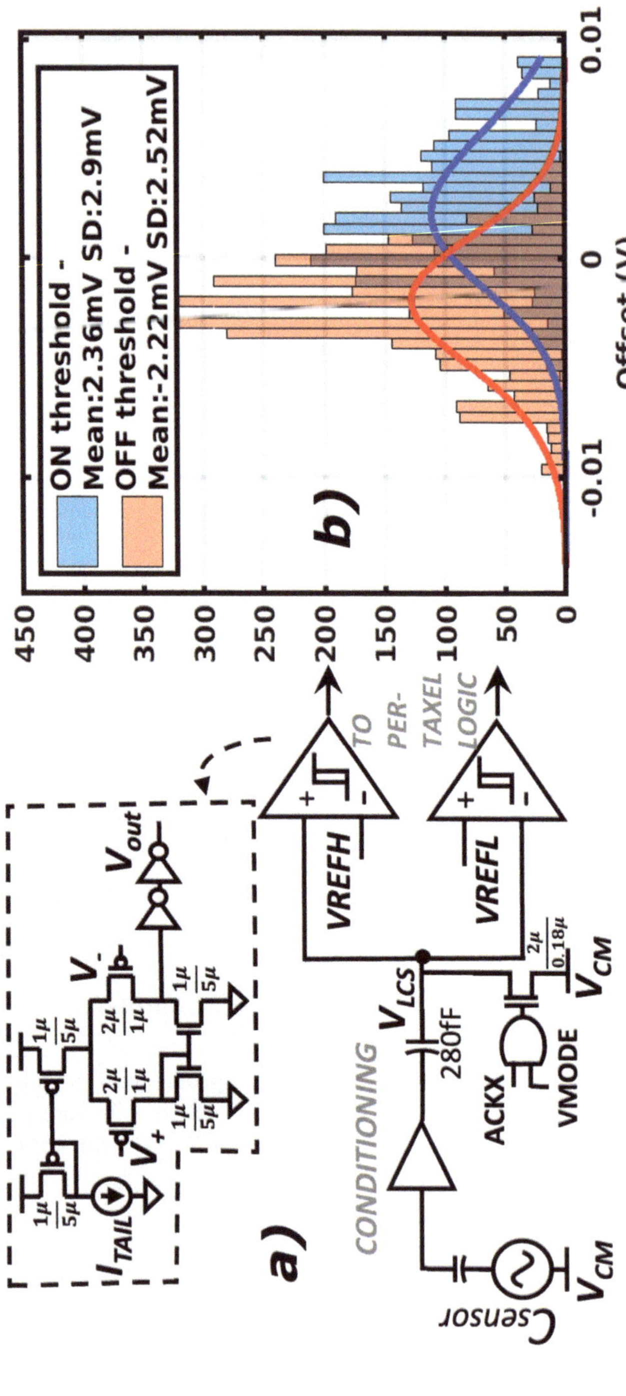

Fig. 4.10 Voltage-mode LC readout circuit: **(a)** schematic; **(b)** ON and OFF threshold offset spread

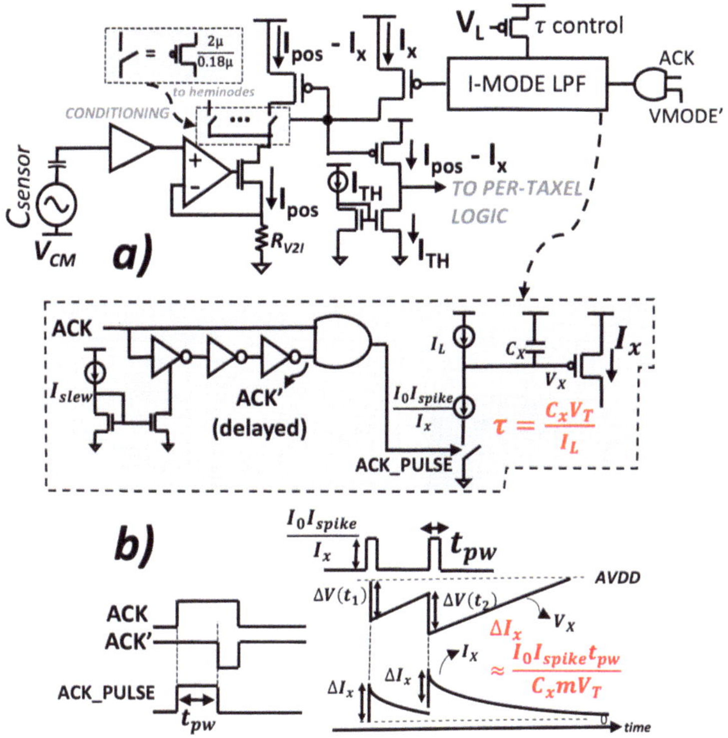

Fig. 4.11 Current-mode neuromorphic-LC readout circuit: (**a**) schematic, and (**b**) transient waveforms illustrating the generation of the exponentially-decaying threshold

Using a PMOS in the subthreshold region (Fig. 4.11a), the linearly increasing V_{GS} bias of $|V_{DD} - V_X|$ results in an exponentially-decreasing I_X with time constant τ :

$$\tau = \frac{C_X \cdot V_T}{I_L} \tag{4.3}$$

The slope of V_X, I_L/C_X, is controlled by changing the gate bias of a PMOS device via V_L (Fig. 4.11a). The increase of the threshold is then achieved by subtracting ΔV_X from V_X:

$$\Delta V_X = \frac{I_0 \cdot I_{SPIKE} \cdot t_{pw}}{C_X \cdot I_X} \tag{4.4}$$

The current $(I_0 \cdot I_{SPIKE})/I_X$ is achieved through the trans-linear principle (Chicca et al., 2014). It is important that the voltage ΔV_X subtracted from V_X is

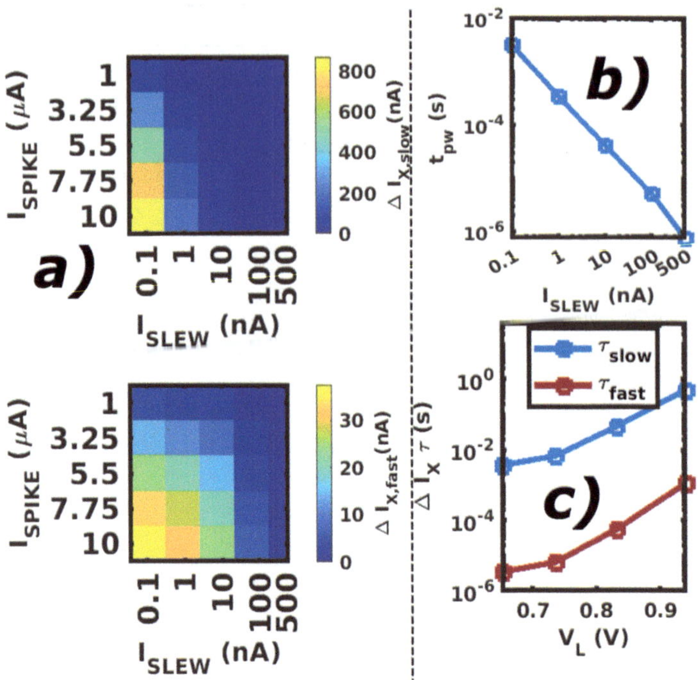

Fig. 4.12 Current-mode neuromorphic-LC readout: (**a**) simulated threshold step size (ΔI_X) for different values of I_{SLEW} and I_{SPIKE}, (**b**) simulated ACK pulse width t_{PW} versus I_{SLEW}, and (**c**) simulated threshold step time constant (τ) versus V_L

an inverse function of I_X to guarantee an additive adaptation to I_X (i.e., $I_{X,final} = I_{X,initial} + \Delta I_X$):

$$I_{X,final} = I_{X,initial} \cdot exp(\frac{\Delta V_X}{mV_T}) \simeq I_{X,initial} \cdot (1 + \frac{\Delta V_X}{mV_T}) \tag{4.5}$$

$$I_{X,final} = I_{X,initial} + \frac{I_0 \cdot I_{SPIKE} \cdot t_{pw}}{C_X \cdot mV_T} \tag{4.6}$$

where m is the subthreshold slope factor.

The threshold step size (ΔI_X) can be controlled through both the ACK pulse width t_{pw} (via I_{SLEW}), and the spike pulse amplitude I_{SPIKE} (Fig. 4.12a and b). The resulting ACK pulse width t_{pw} versus I_{SLEW} (Fig. 4.12c) presents a trade-off between the maximum allowed spiking rate ($1/(2 \cdot t_{pw})$) and the threshold step size (ΔI_X). An identical LPF block with a larger I_L adds to the total $I_X = I_{X,fast} + I_{X,slow}$, resulting in a multi-time-constant exponentially-decaying threshold. The simulated threshold step size decay time constant is shown in Fig. 4.12d, with $I_{SLEW} = 100$ pA and $I_{SPIKE} = 10$ μA for the slow time constant and $I_{SLEW} =$

100 nA and $I_{SPIKEFAST} = 100\,\mu A$ for the fast time constant. A spike is registered when the signal I_{pos}, subtracted with the threshold I_X, exceeds a user-configurable threshold I_{TH} (normally set equal to ΔI_X).

4.6 Experimental Results

Figure 4.13a presents the micrograph of the sensorized chip, while Fig. 4.13b shows the layout of the taxel readout circuit blocks. The *e*-skin chip is fabricated in a 0.18-μm CMOS process, occupying an area of $3.77 \times 5.58\,mm^2$ area with a per-taxel area of $200 \times 200\,\mu m^2$.

4.6.1 *Frontend Characterization*

The test measurement setup is shown in Fig. 4.14a. For chip testing, an on-chip replica frontend chain is used, with its input wire bonded. The replica chain includes a unity-gain buffer at the output ($I_{TAIL} = 10\mu A$) to drive resistive loads. For the

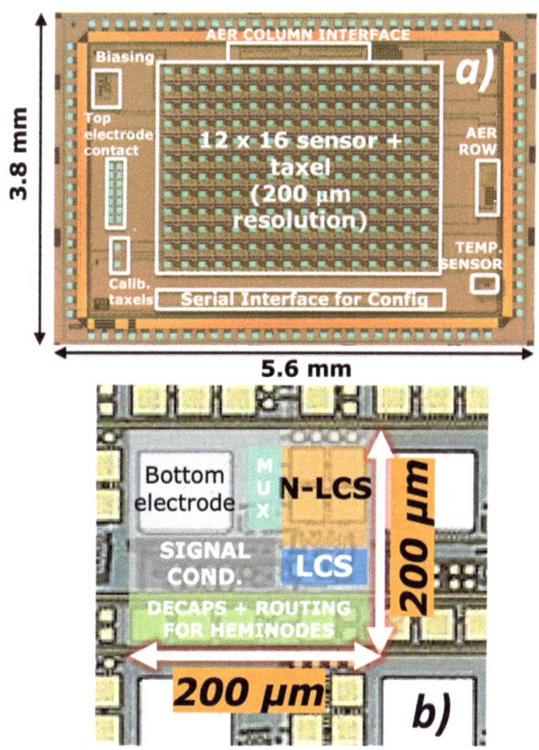

Fig. 4.13 (**a**) INTUITIVE chip micrograph, (**b**) taxel layout (pre-sensor processing) for both the LCS and N-LCS channels. A significant part of the taxel area is due to the bond pad for the sensor connection (*there are no circuits under the pad/bottom electrode*) and the large N-LCS capacitor to achieve a large exponential decay time constant

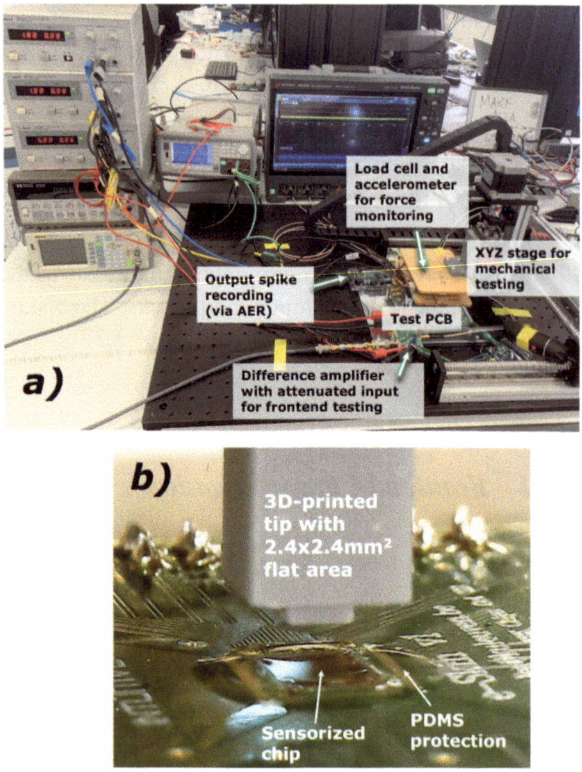

Fig. 4.14 Electrical and mechanical measurement setup: (**a**) the frontend characterization input signal is applied via a commercial difference amplifier that combines both a DC offset and an AC input signal (from Rigol DG3061A). The replica frontend output and input, and the on-chip temperature sensor output are acquired by a Keysight MXR058A. The chip's current consumption is measured by a Keysight B2061A SMU. The output spikes during the experiments are recorded off-chip via an ATMega2560 µC with a software-based AER interface communicating with the chip's AER interface; (**b**) A 3D-printed tip attached to a tri-axial motorized linear stage is used to perform indentation experiments on the sensor chip. A thin layer of PDMS is used for chip protection

frontend characterization, the first-stage charge amplifier (Fig. 4.15) is configured in open loop, with the input signal DC-coupled at V_{CM} and applied directly to the amplifier input. Instead of a more conventional testing methodology, where the charge amplifier is configured in closed loop and an off-chip calibration capacitor together with an AC source is used to model the sensor response, the open-loop testing approach is done because the large R_f from the pseudoresistor can be overcome by any parasitic impedance R_{par} at the wire-bonded input. This then pulls the high-impedance node at the amplifier input into an unknown voltage, shifting it away from the desired value of V_{CM}. Due to the open-loop configuration, the input-referred amplifier offset has a larger effect on the output ($\Delta V_{OUT,DC} \simeq A_{OL1} \cdot V_{offset,ir}$). Thus, a DC level superimposed on the input signal $V_{offset,comp}$, empirically obtained by maximizing the measured open-loop gain, is applied on the AC signal input.

Figure 4.16 shows the measured frequency response of the front-end chain (with the first stage driven directly at the inverting input):

$$A_{OL} = A_{OL1} \cdot A_{CL2} \cdot A_{unity} \tag{4.7}$$

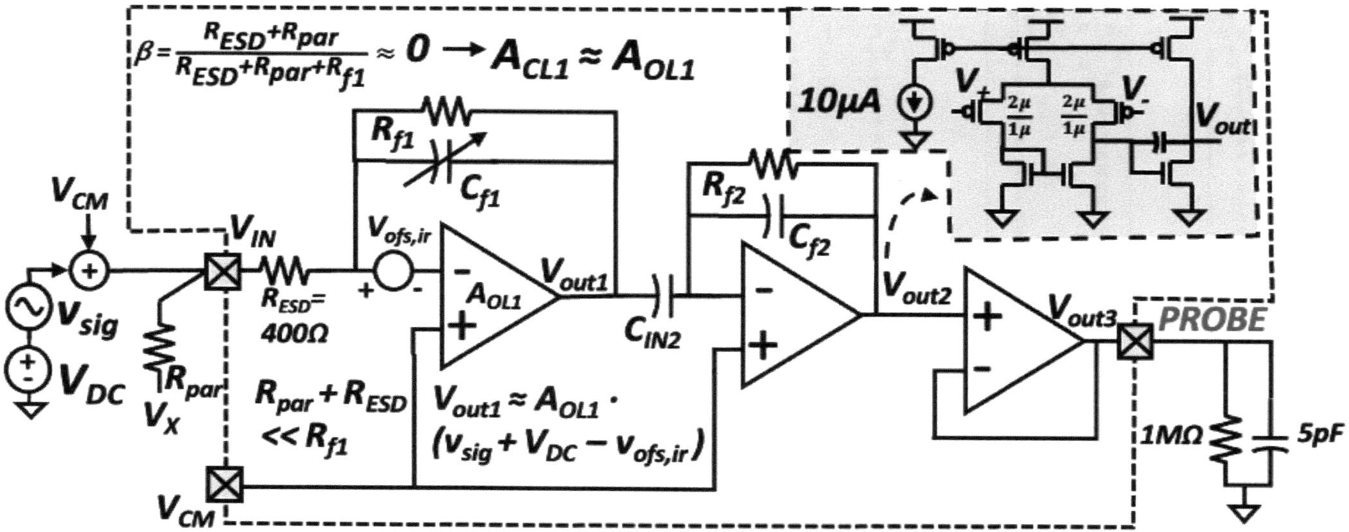

Fig. 4.15 The chip includes a replica frontend channel with an off-chip input and buffered output for characterization. Because the feedback pseudo-resistor R_{f1} presents an ultra-high impedance, any leakage at the bondpad (R_{par}) shifts Vin away from the intended VCM and lowers the amplifier's open-loop gain, confounding the closed-loop measurements. The solution is to apply the input directly to the first stage's non-inverting input and to keep the second stage closed loop; technically the first stage remains closed by the ESD resistor and R_{f1}

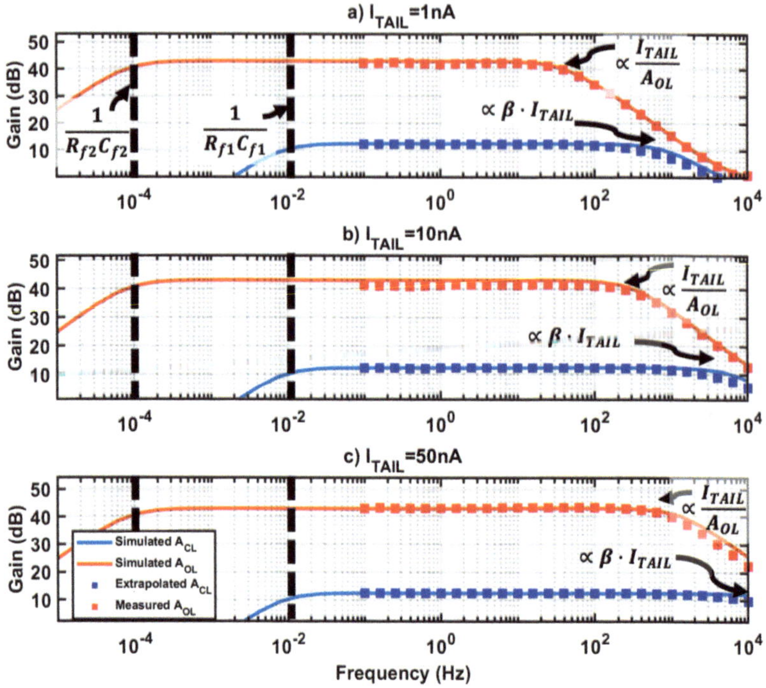

Fig. 4.16 Measured versus simulated open-loop (input signal at the non-inverting input of first stage OTA) and closed-loop gain (input at the top electrode of the sensor) frequency response ($C_{sensor} = 172\ fF$) for: (**a**) $I_{TAIL} = 1\ nA$, (**b**) $I_{TAIL} = 10$ nA, (**c**) $I_{TAIL} = 50$ nA

where A_{CL2} is the 2nd-stage amplifier's closed-loop gain, and A_{unity} is the buffer gain. Both are set to unity.

A_{CL} is then extrapolated from the measured A_{OL1}:

$$A_{CL} = A_{CL1} \cdot A_{CL2} \cdot A_{unity} \tag{4.8}$$

$$A_{CL} = A_{OL1}/(1 + A_{OL1} \cdot \beta). \tag{4.9}$$

Figure 4.17 shows the measured noise PSD. Due to the small input-pair sizing (for reduced area and capacitive loading), the noise PSD is dominated by the flicker noise. For $I_{TAIL} = 1\ nA$, an equivalent input-referred charge resolution of 0.12 fC_{RMS} is achieved for a bandwidth from 1 Hz to 10 kHz:

$$Q_{noise-in,RMS} = \sqrt{(V_{noise-in,RMS})^2 \cdot (C_{sensor} + A_{OL1} \cdot C_{f1})^2} \tag{4.10}$$

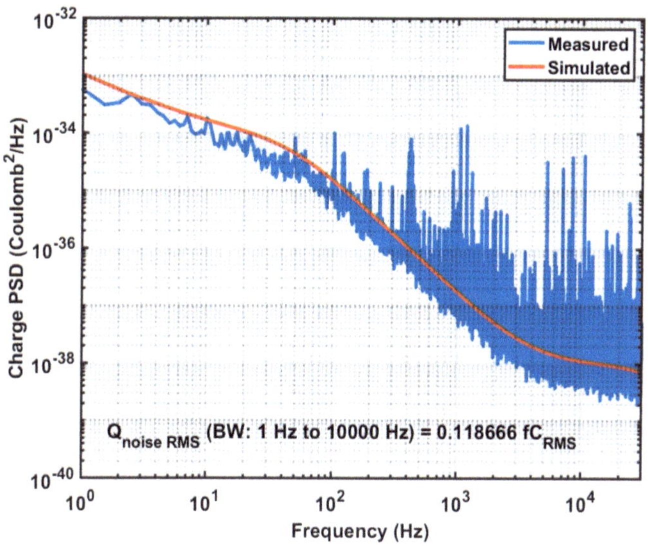

Fig. 4.17 Measured versus simulated noise power spectral density for $I_{TAIL} = 1$ nA ($C_{sensor} =$ 172 fF, $V_{ir-noise} = 24.76\ \mu V_{RMS}$)

Table 4.1 shows that the per-taxel sensor front-end design attains a state-of-the-art performance in charge resolution (270×reduction), area (10×smaller), and power consumption (>30×lower), while having the highest reported number of channels integrated on-chip.

4.6.2 *Taxel Mismatch Characterization*

To characterize the taxel mismatch, including the sensor variability, square pulses of 0.1 Hz are applied to the sensor array's top electrode while the output spikes are recorded. This method is used instead of actual mechanical indentation to eliminate the errors due to the mechanical setup (*tip roughness, chip horizontal flatness, etc*). Figure 4.18a shows the spike train output, while Fig. 4.18b shows the histogram of the number of events per taxel per edge (i.e., the beginning and end of an indentation contact).

A systematic difference resulting in a larger amount of OFF spikes over ON spikes per period is the result of clock feedthrough (Fig. 4.19). The clock's falling edge couples into the high-impedance node, V_{LCS}, which is periodically reset to VCM. This results in a negative shift (ΔV) in the voltage at V_{LCS} (shortly after reset), pushing it closer to V_{CM}—LSB, effectively shrinking the OFF-spike level-crossing window. This results in a systematic bias towards generating more OFF spikes. The inverse effect can be seen on the ON-spike level-crossing window.

Table 4.1 Comparison with state-of-the-art *e*-skin solutions

Reference		Lee et al. (2019)	Rasouli et al. (2018)	Sankar et al. (2022)	Rongala et al. (2019)	Schmitz et al. (2019)	Murray et al. (2020)	This work
Sensors	Type	Piezoresistive	Piezoresistive	Piezoresistive	MEMS	Piezoelectric	Piezoelectric	**Piezoelectric**
	Count	80	6	9	4	13	1	**192 (integrated on chip)**
	Pitch	>1 mm[a]	1 mm	2.5 mm	Not specified	>10 mm[a]	N/A	**0.2 mm**
Readout	Technology	COTS	COTS	COTS	COTS	130 nm	65 nm	**180 nm**
	Power consumption	System: 560 mW Per-taxel: 7 mW	Not specified	Not specified	Readout: 42 mW	System:93.5μW Per-channel:7μW	System:Not specified Per-channel:112 nW	**System: 75μW[b] LCS/N-LCS:12.33/53 nW[c]**
	Power supply	3.3 V	3.3 V	Not specified	5 V	2.5 V	2.5 V	**Taxel + AER:1V I/O:5V**
	Sensor data representation	Spikes (software)	Spikes (software)	Spikes (software)	Spikes (software)	Frames	Frames[d]	**Spikes (on-chip)**
	# of On-chip AFE channels	N/A				13	1	**192**
	Per-channel AFE chip area	N/A				0.24 mm^2	0.0277 mm^2	**0.0027 mm^2**
	AFE charge resolution	N/A				0.1pC	0.1pC	**0.000335pC[e] 0.000444pC[f]**
	ADC resolution	10b	12b	10b	24b	10b	None[d]	**3–5b[g]**
	Acquisition datarate[h]	8 MHz (parallel)	21.6 kHz (sequential)	130 kHz (parallel)	146 kHz (parallel)	16.25 MHz (parallel)	N/A	**<2.235 kevents/s[i] <93.1 events/s[j] (event-driven, spike rate)**

Processing	Classifier	KNN,MLP	ELM	SVM	SNN+KNN	None	None	**SNN+Linear SVM**
	Preprocessing	ISI	None	ISI	None	N/A	N/A	**Texture/Flutter:None**[k] **Indent Period:None**[k,m]
	Classification accuracy	Curvature(3-class):97% Grating(6-class):88%	Texture(10-class):92%	Texture(13-class):97.12%	Texture(10-class):94.2%	N/A	N/A	**Texture(6-class):97.08%**[k] **Flutter(6-class):99.17%**[k,m] **Indent Period(5-class):95.5%**[l]

[a]Estimated
[b]0–62.5k event/s,
[c]$V_{GATE} = 0.245\,\text{V}$
[d]Frontend only
[e]Input-referred charge noise p–p (BW=1 Hz–10 kHz, $C_{sensor} = 172\,\text{fF}$, $V_{ir-noise} = 70.03\,\mu\text{V}_{pp}$)
[f]Input-referred charge noise at $C_{sensor} = 1.72\,\text{pF}$
[g]Level-crossing window
[h]Datarate= $\#bits \cdot f_{sample} \cdot \#sensors$
[i]Texture scan (192 taxels)
[j]Indentation (<10 Hz)
[k]Model sensor input
[l]Integrated sensors
[m]Amplitude-invariant
For a fair comparison of the total neuron count, the N-LCS implementation (in bold) outperforms the classification accuracies of all other implementations

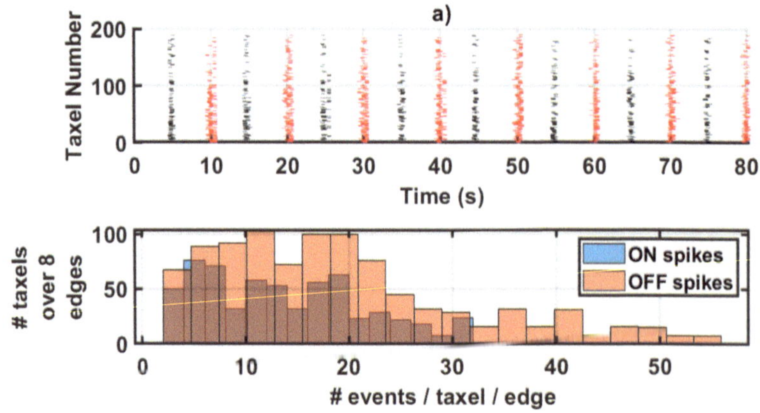

Fig. 4.18 Taxel mismatch after applying 0.1 Hz pulses on the top electrode: (**a**) spike trains, (**b**) histogram of the number of events per taxel per edge over 8 ON/OFF edges

Fig. 4.19 The systematic bias towards generating more OFF spikes is due to the clock feedthrough via the reset switch on the LC-ADC channel

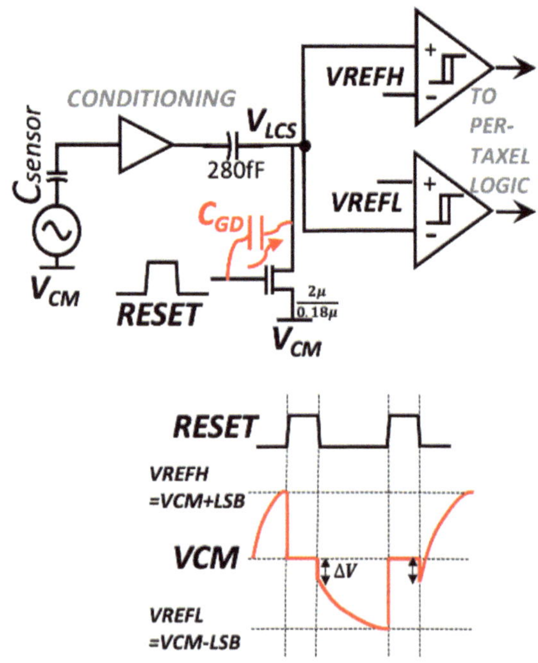

While circuit techniques can be employed to reduce the clock feedthrough, this would introduce additional area and power overhead, which is not aceeptable for dense per-taxel readout. However, we have shown in Sect. 3.1.2 that the systematic mismatch between the number of ON and OFF spikes does not affect the SNN's ability to discriminate the tactile stimuli, allowing to achieve excellent classification accuracies for texture and flutter experiments. Furthermore, by adjusting the voltage references V_{REFH} and V_{REFL} to compensate for the change in the level-crossing window, the systematic offset between the number of ON and OFF spikes can be reduced, as well.

4.6.3 Spike Output Classification

To test the chip's spike generation, the taxel input is emulated through a 16-channel off-chip DAC, while the spike output (from 16 taxels) is read and stored off-chip (Fig. 4.20). The spike addresses are then sequentially re-mapped such that it mimics how an actual tactile stimulus encounters the chip's 12×16 taxels. Figure 4.21a and b present the measured inter-spike interval (ISI), which captures the spike train *noise*, and the measured texture waveform after signal conditioning, respectively.

The tactile stimuli classification using the chip's output spikes (Fig. 4.21c) is performed off-line, similar to Sect. 3.1.2, with a 256-IF neuron, single-hidden-layer SNN network. For the SNN network, a time bin width of 10 ms maximizes the classifier performance for both the texture and flutter frequency spike dataset output. The linear SVM classification of the filtered hidden-layer spikes significantly increases the classification accuracy. This method achieves a classification accuracy (across 6k-folds) of 97.1% on the Kylberg texture dataset (Alea et al., 2022; Kylberg, 2011) and of 99.2% on the flutter (0.5–15 Hz) dataset.

To demonstrate the tactile stimuli encoding of the *e*-skin chip with the integrated PVDF sensors, indentation pulses with varying frequencies are applied $f_{PULSE} = \{10, 5, 2.5, 1, 0.5\}$ Hz. The orthonormal indentation is applied by a 3D-printed mechanical tip with a flat bottom surface ($2.4 \times 2.4\,\text{mm}^2$) attached to a motorized linear stage capable of movement in the *XYZ* directions (Fig. 4.14). Since the temporal information is of more interest than the spatial information in the indentation frequency experiment, it was sufficient to record the population spike output of the entire chip without the localized spatial per-sensor information. The tip area used for indentation is therefore sized such that it fits the sensor active area. The chip's output spikes are then recorded and classified off-chip with an SNN+SVM model, as described in Sect. 3.1.2, achieving a classification accuracy of $95.5\pm4.12\%$ across 6k-folds. The output spike trains are truncated to a maximum of 1600 spikes across all recordings to eliminate the classification bias based only on spike counts. The $<$10-s duration spike trains for all recordings are binned into 0.5-s windows, while the SVM rolling window averaging size is set to 20.

Table 4.1 shows that compared to prior *e*-skin art that performs tactile stimuli classification, this work achieves around 100–7000$\times$reduction in the system power

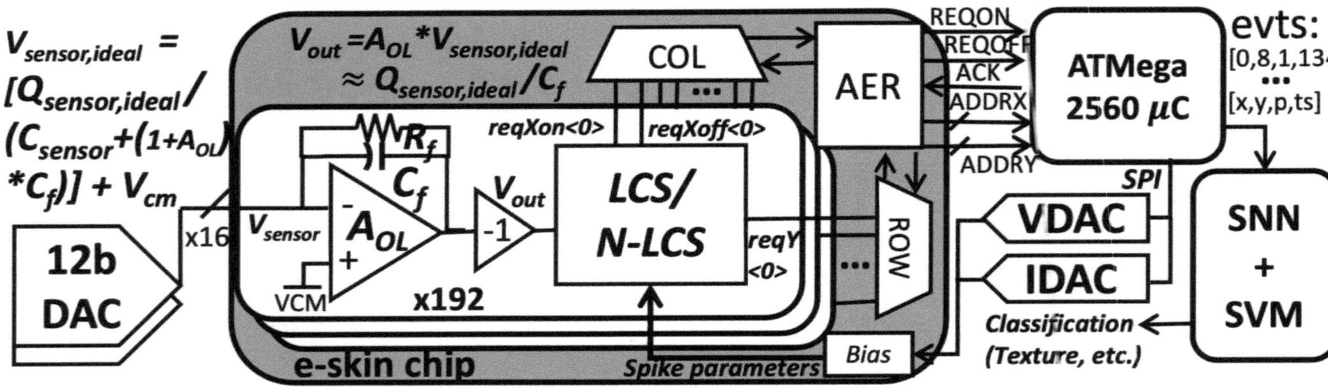

Fig. 4.20 Multi-channel chip testing with classification in offline SNN+SVM software. The expected sensor charge signal as input to the chip is modeled as a voltage signal generated by a 16-channel 12-bit DAC

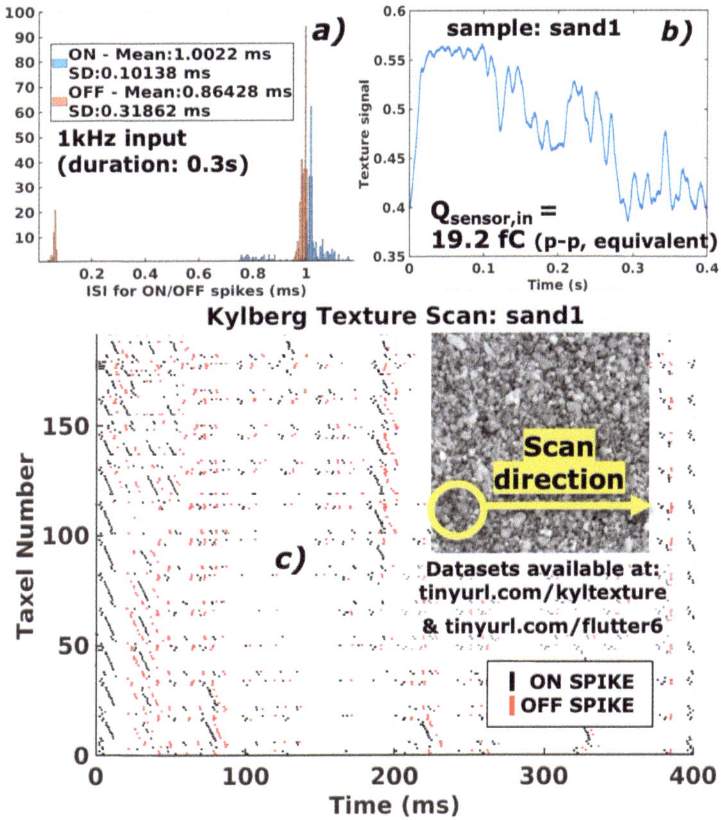

Fig. 4.21 Measurement results: (**a**) Inter-spike interval (ISI) (1 spike/period) showing the spike train noise across 0.3 s. (**b**) Texture signal captured at the output of the replica frontend chain using the taxel #0 model sensor signal as input (C_{sensor} = 172 fF). (**c**) Output spikes induced during texture scan across 192 taxels (shown: *canvas1* sample of the *Kylberg* texture dataset with the scan direction and location highlighted)

consumption and more than 5 orders of magnitude reduction in the per-taxel power consumption, while enhancing the spatial resolution by 5×to 0.2 mm and doubling the sensor count.

4.6.4 Classification Accuracy: LCS Versus N-LCS

To compare the classification accuracy of spikes generated from both the LCS and the N-LCS taxels, spike outputs from 8 taxels with both readout architecures are recorded. The chip's LCS spiking output is obtained with the level-crossing window set at $N = \{2, 3, 4, 5\}$ $bits$, while the N-LCS output is obtained with the threshold

Table 4.2 Comparison of classification accuracy between LCS and N-LCS spiking outputs

Readout architecture	Hidden layer count	Total neuron count	Spiking parameters	Spike rate (Hz)	Classification accuracy[a]
LCS	234	256	$N = 4$ bits	6.13×10^3	$73.3 \pm 17.5\%$
LCS	242	264[b]	$N = 4$ bits	6.13×10^3	$88.3 \pm 7.5\%$[b]
LCS (OFF spikes discarded)	242	256	$N = 4$ bits	2.73×10^3	$75 \pm 12.2\%$
LCS (OFF spikes merged)	242	256	$N = 4$ bits	6.13×10^3	$70 \pm 17.9\%$
N-LCS	**242**	**256**	$\boldsymbol{\tau = 203}$ **ms**	$\mathbf{2.5 \times 10^3}$	$\mathbf{81.7 \pm 7.5\%}$

[a]A subset of the *Kylberg* texture dataset (classes = 6; trials = 10) is used as input
[b]Total neuron count exceeds the 256-neuron limit, making the model unfairly larger than the rest of the comparison
For a fair comparison of the total neuron count, the N-LCS implementation (in bold) outperforms the classification accuracies of all other implementations

decay constant set at $\tau_{equivalent} = \{13, 49, 203, 646\}$ ms. The total spike rate increases with N and decreases with increasing $\tau_{equivalent}$. For a fair comparison between the two spiking readout architectures, a neuron-count-constrained SNN ($neurons = 256$) is used for the hidden-layer count ($HL_{neuron} = 256 - channel_{size} \cdot x_{size} \cdot y_{size} - class_{size}$), where $x_{size} \cdot y_{size}$ is the number of taxel elements, and $class_{size}$ is the number of output classes (of texture). Due to the reduced channel size of the N-LCS spike trains, the N-LCS channels have a slightly larger hidden-layer neuron count at 242 compared to the LCS at 234. Therefore, the classification accuracy for LCS taxels using 242 hidden-layer neurons is also shown to illustrate the effect of this larger hidden layer. Table 4.2 summarizes the peak classification accuracies at the obtained optimum values of the spiking parameters.

Key takeaways from the comparison in Table 4.2 are the following:

- for both LCS and N-LCS taxels, there exists an optimum combination of spike-encoding parameters that maximize the classification accuracy. As seen in Alea et al. (2022), too high spike rates result in a reduced classification accuracy due to possible over-fitting in the training;
- for the LCS taxel, discarding the OFF spikes or treating them as ON spikes in exchange for a larger hidden layer (242) results in a marginally-similar classification accuracy;
- notably, the N-LCS taxel outperforms the LCS across all combinations of encoding parameters, except when the LCS channel is trained with a larger hidden layer of 242.

This difference is likely due to the increased amount of *temporal information* captured by the exponentially-decaying threshold, as opposed to the fixed threshold in the LCS taxel, especially during intermediate periods where the input changes slowly. Interestingly, the N-LCS spiking output is classified better despite generally having a lower spike rate than the LCS outputs. The SNN network's behavior likely results from its preference for not only the specific spiking rates across larger time windows, but also for the distinct temporal patterns of spike occurrences within smaller intermediate windows.

4.7 Conclusion

This chapter has presented the first demonstration of an on-chip per-taxel sensor-to-spike encoding *e*-skin with embedded CRF processing at fingertip-mimicking high taxel density with unprecedented spatial resolution of 200 μm. This INTUITIVE prototype is enabled by the integration of a 12×16 polyvinylidene fluoride (PVDF)-based piezoelectric sensor array on a 0.18-μm CMOS readout chip with dual per-taxel spiking readout channels. The *Spiking Neural Network* (SNN)-based classification performance has been shown to be relatively *insensitive* to the circuit non-idealities, allowing to *relax* the readout circuit design targets, in favor of a reduced area and power consumption, fulfilling the tight requirements for per-taxel readout. The experimental results have confirmed that SNN-based classification of the chip's spatiotemporal spiking output for input tactile stimuli such as texture, flutter, and indentation frequency achieves excellent accuracies, even when using only a small 256-neuron SNN classifier, a low equivalent spike encoding resolution of 3-5 bits, and a sub-Nyquist population spiking rate. Compared to prior *e*-skin art that performs tactile stimuli classification, this work achieves around 2 orders of magnitude reduction in the system power consumption and more than 5 orders of magnitude reduction in the per-taxel power consumption, while enhancing the spatial resolution by 5×to 0.2 mm and doubling the sensor count to 192. Finally, a comparison of the texture classification accuracies between the two on-chip spike encoder outputs has shown that the proposed neuromorphic level-crossing sampling (N-LCS) architecture with a decaying threshold outperforms the conventional bipolar level-crossing sampling (LCS) architecture with fixed threshold.

The INTUITIVE approach has been proven to be highly effective for fine, localized texture sensing, much like the functioning of human fingertips. Human skin, however, performs additional sensing tasks beyond texture perception, most notably slip detection, which requires large-area, multi-modal sensing across both the normal and shear dimensions. Implementing such large-scale functionality in CMOS technologies becomes cost-prohibitive and impractical, calling for a different physical substrate. Moreover, the INTUITIVE system's reliance on piezoelectric sensors limits its response to dynamic normal forces (i.e., vibrations). The DERMIS approach to be presented in the next chapter therefore complements the INTUITIVE system by addressing these missing capabilities, extending the sensing to static and shear forces over broader regions. The design in the next chapter address the limitations of the INTUITIVE approach.

References

Alea, M. D., Safa, A., Assche, J. V., & Gielen, G. G. E. (2022). Power-efficient and accurate texture sensing using spiking readouts for high-density e-skins. In *2022 IEEE biomedical circuits and systems conference (BioCAS)* (pp. 359–363).

Chicca, E., Stefanini, F., Bartolozzi, C., & Indiveri, G. (2014). Neuromorphic electronic circuits for building autonomous cognitive systems. *Proceedings of the IEEE, 102*(9), 1367–1388.

Dahiya, R. S., Adami, A., Pinna, L., Collini, C., Valle, M., & Lorenzelli, L. (2014). Tactile sensing chips with posfet array and integrated interface electronics. *IEEE Sensors Journal, 14*(10), 3448–3457.

Guglielmi, E., Toso, F., Zanetto, F., Sciortino, G., Mesri, A., Sampietro, M., & Ferrari, G. (2020). High-value tunable pseudo-resistors design. *IEEE Journal of Solid-State Circuits, 55*(8), 2094–2105.

Kylberg, G. (2011). The kylberg texture dataset v. 1.0. Technical Report 35, Centre for Image Analysis, Swedish University of Agricultural Sciences and Uppsala University, Uppsala, Sept. 2011. Available online at http://www.cb.uu.se/~gustaf/texture/ and https://kylberg.org/datasets/.

Lee, W. W., Tan, Y. J., Yao, H., Li, S., See, H., Hon, M., Ng, K., Xiong, B., Ho, J., &Tee, B. (2019). A neuro-inspired artificial peripheral nervous system for scalable electronic skins. *Science Robotics, 4*, eaax2198.

Lesniak, D. R., Marshall, K. L., Wellnitz, S. A., Jenkins, B. A., Baba, Y., Rasband, M. N., Gerling, G. J., & Lumpkin, E. A. (2014). Computation identifies structural features that govern neuronal firing properties in slowly adapting touch receptors. *eLife, 3*, e01488.

Marisa, T., Niederhauser, T., Haeberlin, A., Wildhaber, R. A., Vogel, R., Goette, J., & Jacomet, M. (2017). Pseudo asynchronous level crossing adc for ecg signal acquisition. *IEEE Transactions on Biomedical Circuits and Systems, 11*(2), 267–278.

Mensi, S., Naud, R., Pozzorini, C., Avermann, M., Petersen, C. C., & Gerstner, W. (2012). Parameter extraction and classification of three cortical neuron types reveals two distinct adaptation mechanisms. *Journal of Neurophysiology, 107*(6), 1756–1775.

Murray, S. J., Medinger, J., Schmitz, J., Balkır, S., & Hoffman, M. W. (2020). Toward a low power e-skin interface system on a chip for taxel arrays. In *2020 IEEE international symposium on circuits and systems (ISCAS)* (pp. 1–5).

Pinna, L., Ibrahim, A., & Valle, M. (2017). Interface electronics for tactile sensors based on piezoelectric polymers. *IEEE Sensors Journal, 17*(18), 5937–5947.

Rasouli, M., Chen, Y., Basu, A., Kukreja, S., & Thakor, N. (2018). An extreme learning machine-based neuromorphic tactile sensing system for texture recognition. *IEEE Transactions on Biomedical Circuits and Systems,* 1–13.

Rongala, U. B., Mazzoni, A., Spanne, A., Jörntell, H., & Oddo, C. M. (2019). Cuneate spiking neural network learning to classify naturalistic texture stimuli under varying sensing conditions. *Neural Networks, 123*, 231–246.

Sankar, S., Slepyan, A., Iskarous, M. M., Cheng, W.-Y., DeBrabander, R., Zhang, J., Gupta, A., & Thakor, N. V. (2022). Flexible multilayer tactile sensor on a soft robotic fingertip. In *2022 IEEE sensors, SENSORS 2022 - Conference proceedings*. Proceedings of IEEE Sensors (Vol. 2022-October, pp. 1–4). Institute of Electrical and Electronics Engineers Inc.

Schmitz, J. A., Sherman, J. M., Hansen, S., Murray, S. J., Balkır, S., & Hoffman, M. W. (2019). A low-power, single-chip electronic skin interface for prosthetic applications. *IEEE Transactions on Biomedical Circuits and Systems, 13*(6), 1186–1200.

Weber, A. I., Saal, H. P., Lieber, J. D., Cheng, J.-W., Manfredi, L. R., III, J. F. D., & Bensmaia, S. J. (2013). Spatial and temporal codes mediate the tactile perception of natural textures. *Proceedings of the National Academy of Sciences of the United States of America 110*(42), 17107–17112.

Chapter 5
DERMIS: A Fully Integrated Large-Area Grasp-State-Adaptive Tactile Sensor System on a-IGZO TFT

Abstract Chapter 5 presents DERMIS, a fully integrated large-area *e*-skin pathway for low-latency, high-resolution tactile slip sensing in flexible thin-film technology (TFT), combining miniaturized shear/normal force sensors with reconfigurable per-taxel readout and local processing. A floating-electrode differential capacitive sensor enables independent shear and normal force measurement for friction-based slip detection without bulky force/torque sensors, and is optimized for TFT substrate integration. Implemented in a unipolar 0.6-μm a-IGZO TFT process, the fully differential, switch-phase reconfigurable, ratiometric frontend eliminates dedicated charge amplifiers and supports grasp-state-adaptive extraction of shear, normal, and raw differential capacitance. An asynchronous-sampling auto-zero technique within the level-crossing converter reduces flicker noise without the white-noise aliasing of synchronous auto-zero, enabling embedded grasp-state-adaptive feature extraction and data compression from population spikes (contact/lift-off events, friction, and slip). The fabricated prototype achieves 72 μW per-taxel power, 6.16 ENOB, 0.36 mm^2 per-taxel area, ~2-mN_{RMS} force resolution, and a 9×4 mm^2 large-area demonstrator pitch-matched to a 2×8 array with 600 μm shear/force spatial resolution, bringing slip sensing toward human-like performance.

Keywords Slip sensing · Thin-film technology · Capacitive readout · Flexible electronics

In the previous chapters, we have discussed the benefits of end-to-end e-skin integration towards achieving more human-like tactile sensing for future next-generation tactile systems. While a record-density fully-integrated Si-CMOS-based *e*-skin system (INTUITIVE) has been demonstrated in the previous chapter, the large-area sensing requirement for slip detection requires the implementation of the system on a different type of substrate. Human skin performs tasks beyond texture perception—most notably slip detection—which require large-area, multimodal sensing in both normal and shear directions. Such large-scale sensing is impractical and costly in CMOS, and the INTUITIVE system's piezoelectric sensors only respond to dynamic normal forces. The DERMIS approach complements

M. D. Alea, G. Gielen, *Chips That Feel*, Analog Circuits and Signal Processing,
https://doi.org/10.1007/978-3-032-26584-5_5

INTUITIVE by adding static and shear- and normal-force sensing over larger areas. This chapter explains how our work addresses these limitations.

In Sect. 3.2.2, we have highlighted how prior slip-detection techniques using optical- and vibration-based sensors result in excessively-large acquisition data rates and complicated upstream slip detection algorithms. Friction-based approaches, although requiring simpler algorithms, historically have used bulky and expensive multi-axial sensors. Furthermore, prior work only detects slip, neglecting to sense other crucial grasp-state-dependent contact parameters.

In this chapter, the design is presented of a large-area fully-integrated tactile sensor system, called DERMIS, implemented in a flexible thin-film transistor (TFT) technology. An end-to-end design strategy—from the sensor to the readout and to the on-chip feature extraction—enables the following innovations:

- a first-of-its-kind, *biologically inspired*, grasp-state-adaptive tactile system achieved through efficient system-level reconfiguration, with
- *miniaturized multi-axial sensors* enabling a shear sensing spatial resolution of 600 μm, and allowing simple friction-based slip detection without relying on bulky force/torque sensors, and
- sensor-pitch-matched *spike-based readout* in a *large-area flexible TFT* technology (Fig. 5.2c), that implements
- analog-domain encoding of these *grasp-state-dependent* features, eliminating the need for complex offline slip-extraction algorithms, and
- embedded state-machine- and population-output-spike-based grasp-state-aware *feature extraction and data compression.*

5.1 Grasp-state-adaptive Tactile Sensing

Fine grasp-state-dependent force sensing, as performed by human hands during dexterous object manipulation, is likewise crucial for robotic and neuroprosthetic applications (Fig. 5.1). Depending on the current grasp state (Johansson & Flanagan, 2009) (Fig. 5.1b), mechanoreceptors in the human hand sense a variety of contact parameters with high spatial resolution: (a) the *onset/offset* of contact (based on the normal force), (b) *friction* (extracted from the shear force), (c) onset of *lift-off* (based on the change in the shear force) and (d) *slip* (based on the normal and shear forces). When manipulating objects, the brain utilizes sensory predictions and tactile afferent signals to adapt its motor response across sequential action phases, monitoring contact events that trigger rapid corrective actions (Johansson & Flanagan, 2009). Therefore, adding grasp-state-aware force-sensing capabilities into tactile sensors is crucial in real-world grasp applications.

Current tactile sensors (Fig. 5.2a and b) offer low spatial resolution, rely on hardware-expensive slip-detection algorithms, and, as discussed above, cannot reconfigure dynamically with the grasp state. To address this, we introduce our DERMIS *end-to-end* design approach (Van Assche et al., 2023), which leverages

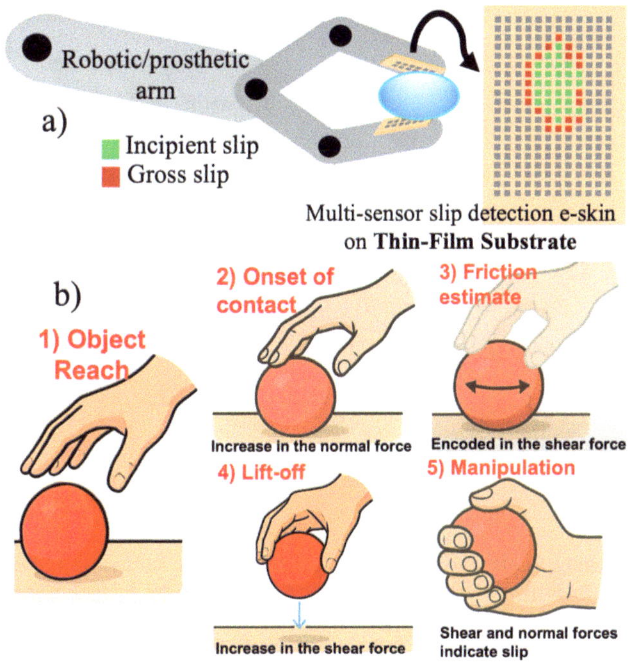

Fig. 5.1 (**a**) The high-spatial-resolution tactile sensor allows detecting localized slip, discriminating between incipient and gross slips. Object slips are typically characterized by an inner *stick region* (*in green*) where incipient (partial) slip happens, and an outer region (*in red*) where gross slip is seen. (**b**) Different grasp states during object manipulation. The illustrative image was generated with assistance from OpenAI's ChatGPT image generation tool

the flexibility of a full-custom top-down methodology, with the dense integration enabled by VLSI technologies, the system incorporates miniaturized tactile sensors with a shear spatial resolution of 600 μm, and allowing simple friction-based slip detection without relying on bulky force/torque sensors. Through a state-machine-based reconfiguration of the readout, depending on the current grasp state, measurement of not only *slip*, but also additional contact parameters essential for robust contact and slip detection, including *friction* and the *onset/offset* of contact and lift-off—capabilities is possible. This is a feature that has not been available in existing tactile sensors.

5.2 TFT-based Tactile Sensing System

While a modern fully-integrated Si-CMOS-based high-density *e*-skin system has been demonstrated in the previous chapter (Alea et al., 2024), the large-area sensing requirement for slip detection demands implementing the system on a different

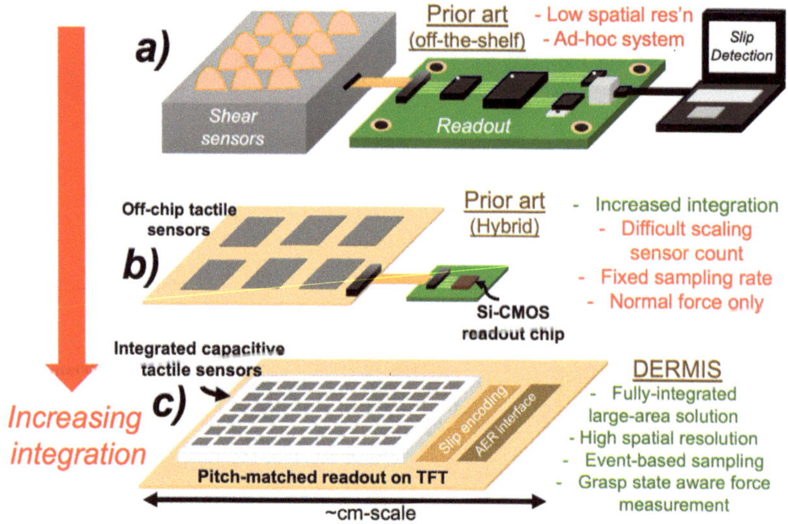

Fig. 5.2 (**a**) Existing tactile sensing systems lack the level of integration to allow for power-efficient and low-latency slip detection. (**b**) Hybrid approaches are limited in scalability. (**c**) To match the high spatial resolution and the fast localized slip detection capability of human hands, next-generation large-area slip detection systems can leverage the integration potential of thin-film transistor (TFT) technology, as we do in our DERMIS design

type of substrate. As described in Chap. 2, previous large-area flexible *e*-skins are implemented on exotic substrates (Kim et al., 2018; Liu et al., 2022) and, thus, lack the monolithic integration capabilities to implement the more complex embedded readout and processing needed. To achieve the various requirements and capabilities required in next-generation integrated slip detection and grasp-state-adaptive tactile sensing systems, a thin-film transistor (TFT) technology-based approach is an attractive alternative (Fig. 5.2c), and will be used in the design presented in this chapter.

In contrast to Si-CMOS-based designs, TFT-based solutions are larger in area and are flexible, but they are constrained by the performance of the monolithic devices available. TFT transistors typically have lower carrier mobilities and higher threshold voltages (Zulqarnain & Cantatore, 2021), resulting in larger device sizes and supply voltages. Furthermore, a-IGZO-based designs today are unipolar (i.e., NMOS only). Note that complementary devices (i.e., PMOS) are being developed, but these are not yet available for designing. As a consequence, implementing complex digital computation, possible in Si-CMOS with its dense integration, becomes a challenge in the available TFT technology. TFT devices also suffer from larger flicker noise, large mismatch, and fabrication defects (Aygun et al., 2019). Therefore, large-area TFT-based designs for embedded slip detection benefit rather from largely analog-based algorithms with simple digital blocks, as proposed in this chapter.

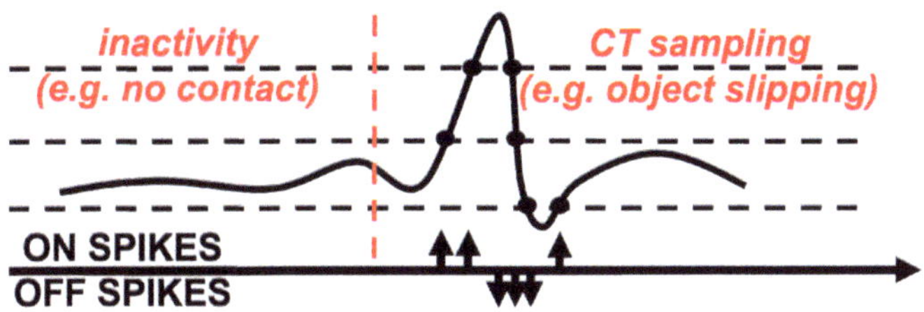

Fig. 5.3 Event-driven readout scales its signal sampling rate based upon the instantaneous bandwidth of the tactile input signal, converting the continuous-time sensor signal into an asynchronous spike train

Table 5.1 Comparison of the DERMIS system with state-of-the-art slip-detecting tactile sensor systems

	Romeo et al. (2017)	Fernández et al. (2014)	Cheng et al. (2010)	Su et al. (2015)	Tremblay and Cutkosky (1993); Leslie et al. (2023)	**This work** (DERMIS)
Sensor type	Vibration	Vibration	Normal force	Shear + Normal force	Shear + Normal force	**Shear + Normal force**
Slip detection algorithm	Envelope detection	Transforms	Inter-frame cross-correlation	Classifier-based	Friction extraction (post-quantization)	**Friction extraction (pre-quantization)**
Direct slip measure-ment?	No	No	No	No	Yes	**Yes**
TFT Hardware-friendly?	No	No	No	No	No	**Yes**

Our implementation in BOLD

As described in Chap. 3, adaptive event-based signal acquisition (Alea et al., 2022) is beneficial since both fast slip sensing (<10 ms for safe grasping) and sparse spatiotemporal sampling (during periods of inactivity) are required (Fig. 5.3). Our DERMIS fully-integrated solution, therefore, combines capacitive shear sensors with event-based, reconfigurable readout in a large-area TFT technology to enable high-resolution grasp-state-dependent force sensing, including slip detection.

Table 5.1 compares our proposed DERMIS solution to other existing slip-detecting tactile sensor systems. The proposed solution can measure slip directly in a non-compute-intensive manner, and therefore is a viable approach for integration in a digital-logic-constrained TFT process. To allow grasp-state-dependent force sensing, the DERMIS system incorporates the following innovations: (a) miniaturized force sensors with both normal and shear force transductions, (b) a reconfigurable

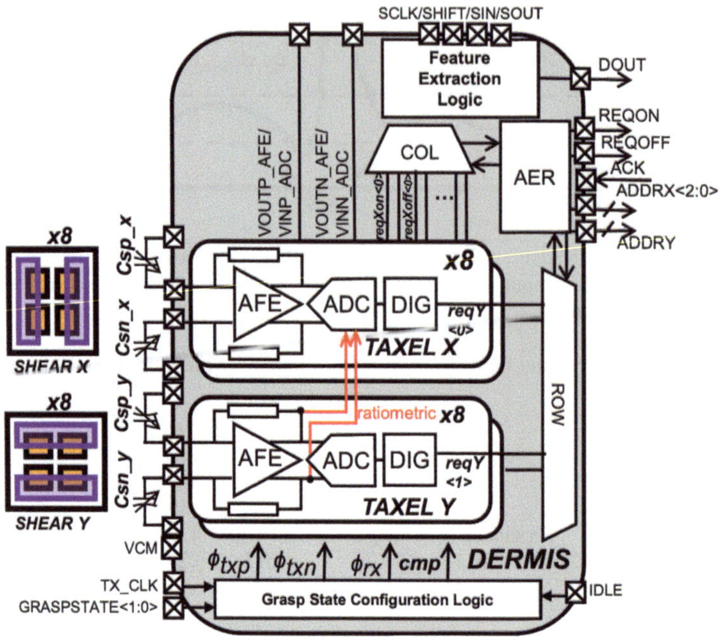

Fig. 5.4 Top-level architecture of the DERMIS *e*-skin system. Separate readouts for the shear sensors in both the X and Y orientations allow the detection of the shear direction. An asynchronous Address Event Representation (AER) interface sends out an output spike train after sensor signal conversion. To extract a measure of the shear force, a ratiometric mode (in *red*) uses the output of a neighboring taxel's analog frontend (AFE) unit as a reference for ADC quantization, essentially performing a division. Orienting sensors in both the X and Y directions allow for rotational slip detection (Castaño-Amorós and Gil, 2023), in addition to just the linear slip. The `GRASPSTATE<1:0>` and the `IDLE` pins are used to reconfigure the readout mode (i.e., select between shear and normal force acquisition), and also to control the state machine in Fig. 5.9

readout to encode the normal and shear forces, and (c) on-chip grasp-state-aware data compression and feature extraction. We will now describe the system in greater detail.

Our proposed implementation focuses on a fully monolithic TFT design (Fig. 5.4).

5.2.1 Sensors

Novel differential capacitive sensors allow for sensing both the normal and shear forces in the *X* and *Y* directions (e.g., by rotating the sensor pattern perpendicularly), enabling grasp-state-dependent measurements to be performed without the bulky form factor of traditional force/torque (F/T) sensors (Romeo & Zollo,

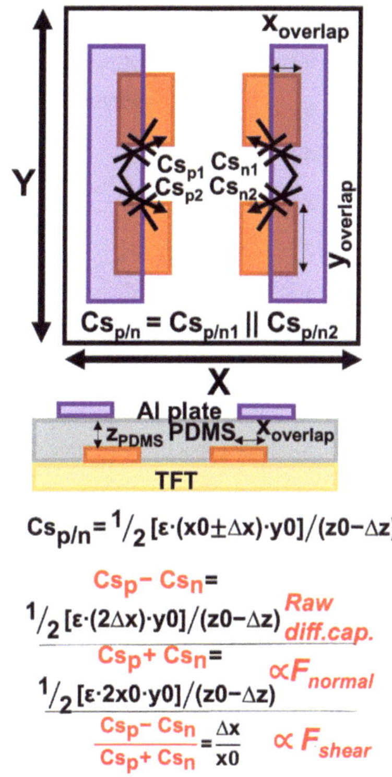

Fig. 5.5 Differential capacitive sensor for both normal and shear force detection, designed to be integrated on top of a TFT substrate with integrated readout electronics. The bottom part gives the equations to extract the normal and shear forces

2020). Conventional capacitive tactile sensors, however, require both top and bottom electrodes, making the top-electrode connections difficult and prone to wear during contact. The proposed design introduces a novel "floating top electrode" architecture that eliminates this issue by routing all electrical connections through the two bottom electrodes, while the middle electrode (see Fig. 3.32) remains floating, forming an equivalent series capacitance. The differential configuration enhances the noise immunity and enables the independent measurement of normal and shear forces, unlike prior shear sensors where shear readings are coupled to the normal loading.

Shear and normal forces can be measured via simple arithmetic operations on the differential capacitance, as will be discussed in detail in Sect. 5.2.2. The capacitive sensor structure used is shown in Fig. 5.5. The capacitive sensors are formed through the overlap of a floating top electrode (Al) and two bottom electrodes (TFT top metal, Al), forming two capacitances in series. The sensors are fabricated on a wafer carrier through a novel sacrificial molding layer approach (Rosa & Kraft, 2024). The fabrication is done by Maria Rosa, our collaborator from the Micro and Nanosystems Group. The structure is then integrated on the TFT substrates through an alignment-based post-processing step after regular TFT chip fabrication.

The dynamic range of the capacitive sensor is modeled in Fig. 5.6, considering changes in both the shear and normal forces, which modulate Δx and Δz,

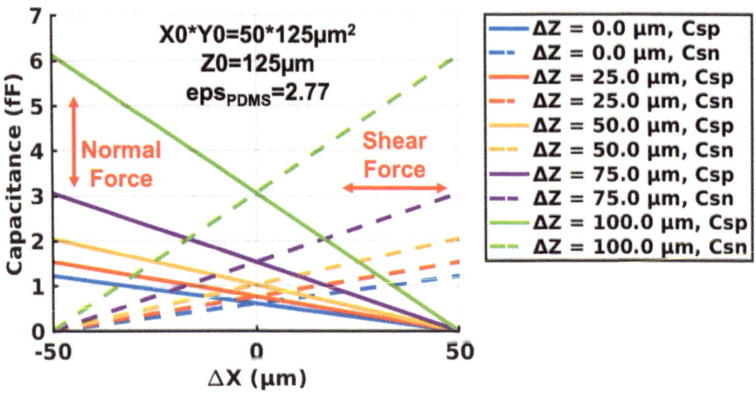

Fig. 5.6 Modeled capacitive dynamic range of the integrated multi-axial tactile sensors with 600-µm pitch

respectively. Note that the per-taxel readout is pitch-matched to the capacitor sensor resolution of 600 µm in our DERMIS design.

5.2.2 Readout Circuit

The differential nature of the signal conditioning improves beyond the INTUITIVE design: (a) by improving the common-mode noise rejection, and (b) by allowing for analog-based arithmetic operations on the differential capacitive sensor readouts, isolating either the normal or shear forces. Leveraging the intrinsic shear- and normal-force detection capabilities of the tactile sensors, the readout system reconfigures itself based on the current grasp state, achieving the same grasp-state-adaptive force measurements as performed in the human somatosensory system. The embedded grasp-state-adaptive feature extraction and data compression logic is implemented in a unipolar a-IGZO TFT process for the first time, utilizing the population output spikes to detect crucial grasp-state-based parameters such as the onset/offset of contact, the onset/offset of lift-off, initial contact friction, and slip.

The designed readout circuit (Fig. 5.7) takes advantage of the versatile nature of the differential capacitive sensors, inherently allowing the extraction of both shear and normal forces through a simple reconfiguration of the phases of the capacitor stimulation voltages. The normal and shear forces are extracted by performing simple arithmetic operations on the raw capacitance values extracted from the sensor (see Fig. 5.5). Such operations can intrinsically be performed on the AFE output before quantization. This approach eliminates the need for any compute-intensive algorithm, which is expensive to implement digitally in a density-limited TFT technology, and would result in large detection latencies to extract such forces after quantization. The normal force can be extracted by applying out-of-phase

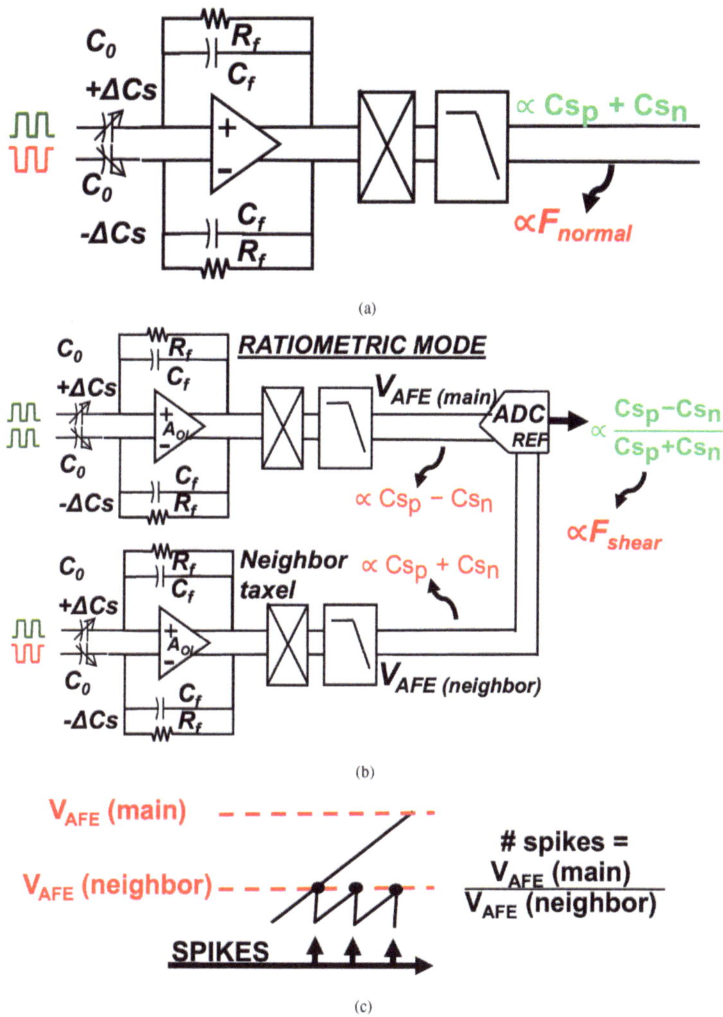

Fig. 5.7 Extracting the normal and shear forces via the DERMIS readout: (**a**) this configuration measures the normal force, while (**b**) this ratiometric configuration measures the shear force, as depicted in (**c**)

stimulation waveforms on Csp and Csn, resulting in a voltage output that is proportional to the sum of the two capacitances after demodulation and filtering (Fig. 5.7a, MODE:FN). A ratiometric mode (Fig. 5.7b and c, MODE:FS) allows to extract the shear force by using the output of a neighboring taxel as the reference for the quantization step of the ADC (essentially performing a division). A low-power mode (MODE:$IDLE$) allows the chopping mechanism to be turned off, permitting a reduced bandwidth and consequently reduced power requirements for the frontend OTA.

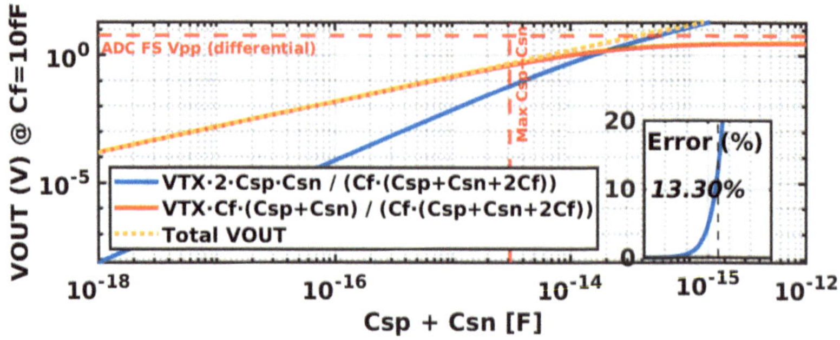

(a)

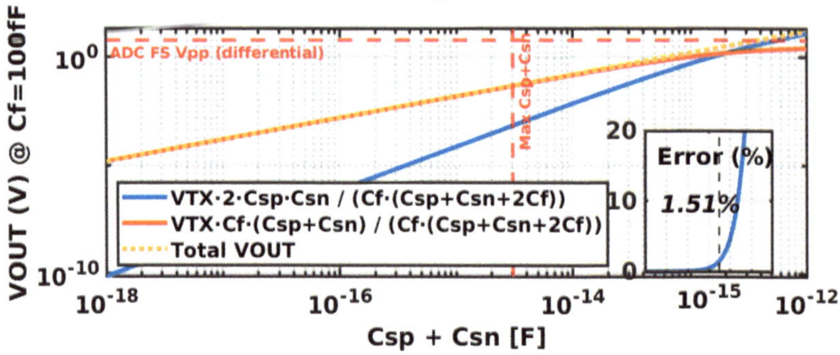

(b)

Fig. 5.8 The nonlinear term in the derived AFE output voltage contributes a small error term (Eq. (5.3)). The figure shows the AFE output versus the normal force (as expressed by *Csp*+*Csn*) for (**a**) *Cf=10 fF* and (**b**) *Cf=100 fF*. The inset plots show the error (%) due to this nonlinear term at the maximum *Csp*+*Csn* value, denoted by the vertical dashed line

The AFE output expressions across different grasp states are summarized in Eqs. (5.2)–(5.6), where V_{TX} is the peak-to-peak voltage of the square-wave input applied to the equivalent sensors top plate (e.g., by treating the two series capacitors as a single equivalent capacitor on one side of the differential sensor), $VOCM$ is the output common-mode voltage (set by the common-mode feedback), and *Cf* the feedback capacitance. In $MODE{:}FN$, the nonlinear term in the AFE output (Eq. (5.3)) is reduced because the sensor exhibits small capacitance variations throughout its dynamic range (Fig. 5.8). The neighbor taxel AFE, when configured in ratiometric mode (MODE:FS), is configured to have a smaller closed-loop gain (i.e., a larger feedback capacitance by factor K) to make sure the input signal exceeds the threshold (Fig. 5.7c and Eq. (5.7)).

$$V_{\mathrm{OUT}}(\mathrm{raw}) = \frac{2V_{\mathrm{TX}}(\beta_2 - \beta_1)}{\beta_1 + \beta_2} + \frac{2V_{\mathrm{OCM}}(\beta_1 - \beta_2)}{\beta_1 + \beta_2} \tag{5.1}$$

$$= \frac{V_{\mathrm{TX}}\left(C_{sp} - C_{sn}\right)}{C_{sp} + C_{sn} + 2C_f} \quad \propto \text{raw diff. cap. value} \tag{5.2}$$

$$\because\ V_{\mathrm{OCM}} = \frac{V_{\mathrm{TX}}}{2}$$

$$\because\ \beta_1 = \frac{C_f}{C_{sp} + C_f} \quad \beta_2 = \frac{C_f}{C_{sn} + C_f}$$

$$V_{\mathrm{OUT}}(\mathrm{FN}) = \frac{V_{\mathrm{TX}}\left[\,2\,C_{sp}C_{sn} + C_f\left(C_{sp} + C_{sn}\right)\right]}{C_f\left(C_{sp} + C_{sn} + 2C_f\right)} \tag{5.3}$$

$$V_{\mathrm{OUT}}(\mathrm{FN}) \approx \frac{V_{\mathrm{TX}}\left(C_{sp} + C_{sn}\right)}{C_{sp} + C_{sn} + 2C_f} \qquad \propto F_{\mathrm{normal}} \tag{5.4}$$

$$V_{\mathrm{OUT}}(\mathrm{FS}) = \frac{V_{\mathrm{OUT}}(\mathrm{raw})}{K\ V_{\mathrm{OUT}}(\mathrm{FN})} \propto \frac{C_{sp} - C_{sn}}{C_{sp} + C_{sn}} \quad \rightarrow \propto F_{\mathrm{shear}} \tag{5.5}$$

$$V_{\mathrm{OUT}}(\mathrm{idle}) = \frac{2V_{\mathrm{OCM}}\,(C_{sn} - C_{sp})}{C_{sp} + C_{sn} + 2C_f} \tag{5.6}$$

$$\because\ C_f' = 4C_f \ \Rightarrow\ K = \frac{1}{4} \tag{5.7}$$

5.2.3 *Embedded Grasp-state-based Feature Extraction*

The human somatosensory system relies on fast *feature extraction* in navigating the various stages of grasping (e.g., fast object slip detection to adjust the grip force). Inspired by this, the DERMIS prototype system performs grasp-state-aware feature extraction while effectively performing on-chip data compression locally to reduce the detection latency. Dexterous object manipulation consists of different grasp states, which involve planning the next action depending on the extracted features at the current grasp state. This evolution is depicted in Fig. 5.9a, while the data compression waveforms are shown in Fig. 5.9b. Table 5.2 summarizes the relevant feature extracted per state. The path highlighted in green depicts the actions performed before active object manipulation (e.g., picking an object from a table), while the path in red depicts post-manipulation movements (e.g., placing the object on a table).

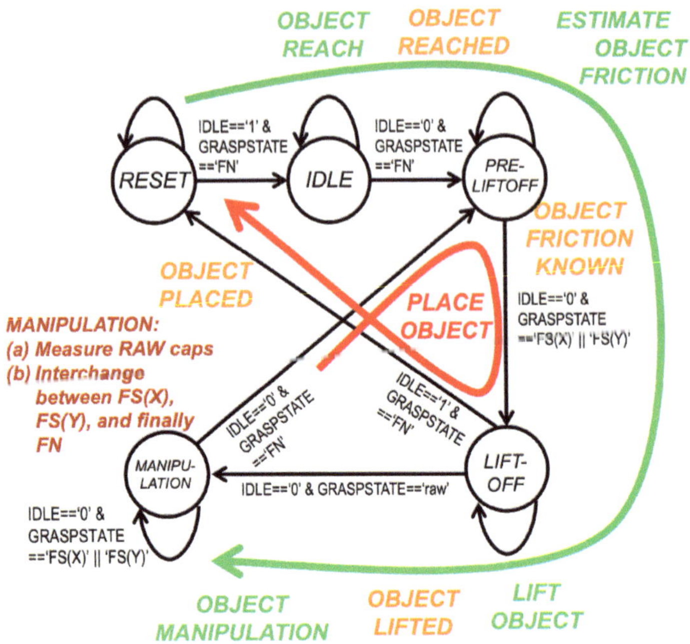

(a)

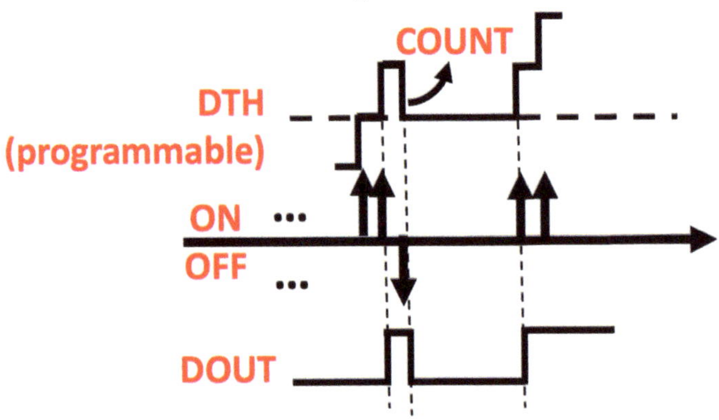

(b)

Fig. 5.9 (**a**) State diagram during the evolution of grasping movements. Similar to humans, grasp manipulation involves planning the next action depending on the extracted feature at the current state. (**b**) Grasp-state-based data compression is achieved by integrating ON and OFF spikes in a counter. When certain programmable thresholds (across grasp states) are crossed, a binary signal is asserted. This signal informs the external controller (e.g., a prosthetic user or a robotic system) that a grasp state has concluded and that it is safe to proceed to the next state

Table 5.2 Force features extracted for different grasp states

STATE MEASURED (DOUT='1')	FORCE FEATURE EXTRACTED	
IDLE	Normal force	Onset of contact
LIFT-OFF	Shear force	Lift-off
MANIPULATION	Normal & Shear force	Slip

5.3 Detailed Circuit Design

5.3.1 System Architecture

The DERMIS TFT readout chip prototype includes a 2×8 taxel array with per-taxel readout, implemented in a Pragmatic unipolar 0.6 μm a-IGZO technology (Pragmatic Semiconductor, 2024). Each per-taxel readout combines a capacitive frontend and an event-based level-crossing (LC) data converter (Fig. 5.10). The frontend comprises a fully differential capacitively-coupled amplifier, with the differential capacitive sensor directly used as the input capacitance, eliminating the need for extra charge amplifiers (Park et al., 2019). The LC converter (Van Assche & Gielen, 2020) asynchronously detects signal changes, using window comparators to generate bipolar spikes encoded as addresses by an on-chip AER interface (Boahen,

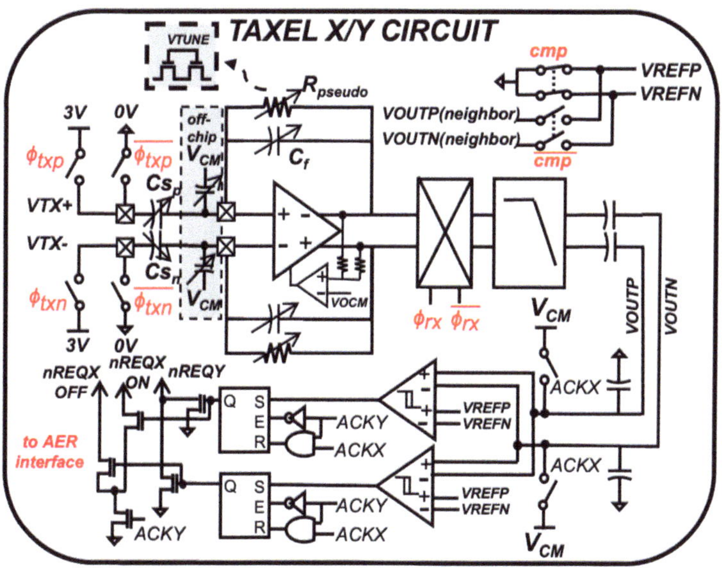

Fig. 5.10 Schematic of the per-taxel readout (2×8 taxels are integrated on-chip) showing the fully-differential AFE and the event-driven LC-ADC. The comparator inputs use a neighboring taxel's AFE outputs during the ratiometric mode

2000). Per-taxel digital blocks, implemented with differential logic for reducing the static power (Çeliker et al., 2024), latch the LC events, which are then arbitrated and transmitted off-chip by the AER interface. This interface, which arbitrates outputs from multiple taxels, uses pseudo-CMOS logic for a good noise margin, trading off for increased static power (Çeliker et al., 2024).

5.3.2 *TFT Design Risks*

TFT designs are prone to several design risks in practice:

- the relatively immature device models and the lack of Monte-Carlo and noise models;
- the relative immaturity of the digital logic libraries;
- the lack of a proven TFT digital design implementation workflow;
- the lack of ESD protection (including diodes), making bench measurements challenging;
- the lack of standard IO cells and bondpads;
- the challenge of wirebonding on thin and flexible TFT substrates.

5.3.3 *AFE OTA*

Despite its tight 0.04 mm^2 area, the fully differential amplifier (Fig. 5.11) achieves an open-loop gain of 30 dB using a "pseudo-PMOS" load implemented with a positive feedback network (Dandekar et al., 2022). The positive feedback effectively

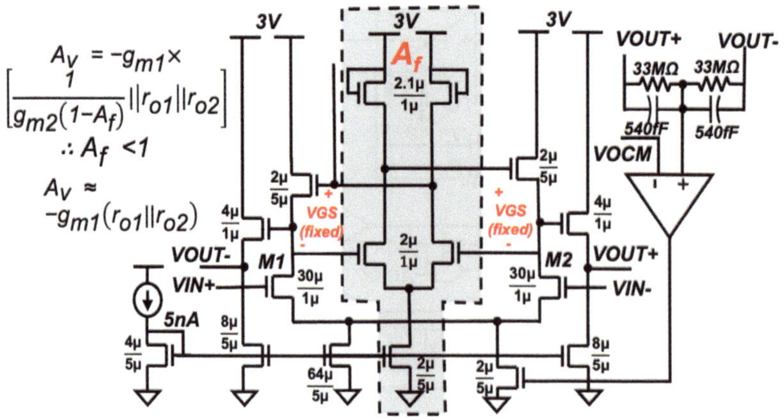

Fig. 5.11 Schematic of the fully-differential AFE OTA, highlighting the pseudo-PMOS load (through positive feedback) to boost the gain

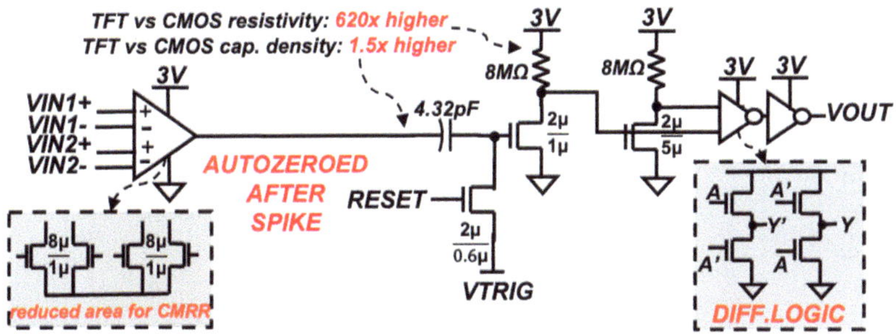

Fig. 5.12 Schematic of the non-uniform-sampling auto-zeroed comparator

keeps the V_{GS} of the NMOS load constant, eliminating its transconductance contribution to the small-signal gain, keeping only the higher intrinsic output impedance. The common-mode feedback (CMFB) amplifier is implemented with an NMOS diode-load OTA.

5.3.4 Auto-zeroed Comparator

Two continuous-time (CT) comparators detect the residue voltage's level crossings (± 1 LSB), each using a 3-stage open-loop amplifier: a dual-differential input preamplifier and two resistive-load common-source (CS) stages (Fig. 5.12). The preamplifier's dual-differential input rejects common-mode noise, including switching artifacts, and allows comparison with a neighboring taxel's differential output in the system's ratiometric mode.

The high-pass LC conversion enables AC coupling of the first AC stage, decoupling its biasing from the preamplifier's DC output (set by the NMOS load gate-to-source voltage), thereby maximizing gain. To move the high-pass corner to low frequencies, this approach requires a large capacitance and a high-impedance path to the desired DC level, V_{TRIG}. Measurement results from the first chip suggests this high-pass corner is significantly moved to higher frequencies, effectively attenuating the input signal, resulting in reduced level crossings. This is possibly due to the higher subthreshold leakage current of the NMOS-based impedance to the DC level. An improvement in the second chip is to replace the high-impedance path with an NMOS switch, that resets every level crossing, effectively resulting in a non-uniform auto-zeroed comparator. The auto-zero effectively attenuates the low-frequency flicker noise, but typically introduces noise folding and elevates the overall noise floor (Rooijers et al., 2022). Because non-uniform sampling is inherently aliasing-free (Tsividis, 2010), this issue is however effectively mitigated. The proposed non-uniform AZ technique has been modeled and simulated. Figure 5.13a shows the flicker noise attenuation without noise folding as compared to a uniform AZ,

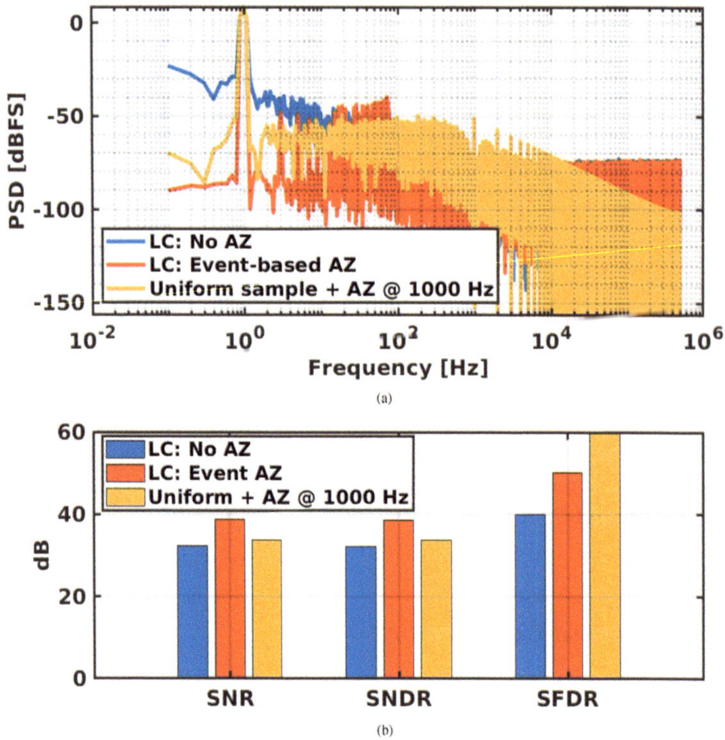

Fig. 5.13 (**a**) Simulated PSD comparing the effect of non-uniform-sampling versus conventional uniform-sampling auto-zeroing. The raw PSD with flicker noise is added for reference. (**b**) The achieved SNDR, SNDR and SFDR across the following: LC-ADC (no auto-zero), LC-ADC (auto-zero), and uniform-sampling auto-zero

while Fig. 5.13b compares the SNR, SNDR, and SFDR, highlighting the benefits of the non-uniform AZ.

An added advantage of the high-pass LCS conversion is that it decouples the first glscs stage's biasing from the preamplifier's DC output (set by the NMOS load V_{GS}), thereby maximizing the gain of the next stage. Large resistive loads bias the CS stages, reducing the power consumption by leveraging the high-resistivity devices available in the TFT technology used.

5.3.5 *Per-taxel Grasp-state-based Control Logic*

The DERMIS per-taxel readout (Fig. 5.10) can be reconfigured to detect grasp-state-dependent forces, as summarized in Table 5.3. This happens by adjusting the phases of the V_{TX} switches ($\phi_{txp/n}$) on the capacitive sensors Csp and Csn,

Table 5.3 Truth table to reconfigure the readout depending on the grasp state (FN, $FS(X)$, $FS(Y)$, raw, and $IDLE$ modes)

TAXEL X				TAXEL Y					
FN	FS(Y)	FS(X)	Raw	FN	FS(X)	FS(Y)	Raw	IDLE	Signal
1	0	1	0	1	0	1	0	X	TX_CLK
1	0	1	0	1	0	1	0	0	ϕ_{txp}
0	1	1	0	0	1	1	0	0	ϕ_{txn}
1	0	1	0	1	0	1	0	1	ϕ_{rx}
1	0	1	1	1	0	1	1	1	cmp

driven to be either V_{TX} (3V) or 0V. The V_{TX} switches are driven with a 6-V gate drive via a level shifter. The chopper demodulation switch gate logic can also be reconfigured ($\phi_{rxp/n}$), and is permanently asserted during the $IDLE$ mode (non-switching mode). The cmp signal is used during ratiometric modes, where it decides whether to take the local AFE outputs or the neighboring taxel's AFE outputs as inputs to its comparator. The on-chip grasp-state-based control logic is implemented with differential logic for reduced static power consumption.

The feedback capacitance, shown in Fig. 5.14, can be configured in two ways: via the `GAINHI` setting and by operating in ratiometric mode (MODE: `FS`). As shown in Fig. 5.7c, spikes are generated whenever the main taxel's AFE output exceeds the neighboring taxel's AFE output (i.e., under ratiometric operation). To guarantee that spikes are generated, the neighboring taxel's AFE output waveform is intentionally attenuated by a factor $K = 1/4$, as shown in Eq. 5.7, by increasing the feedback capacitance 4× from 40 fF to 160 fF in the normal-gain setting (`GAINHI` = '0') and from 10 fF to 40 fF in the high-gain setting (`GAINHI` = '1').

5.4 Experimental Results

5.4.1 Sensor Measurements

To test the sensor fabrication methods and validate the sensing principles, a protoype capacitive sensor with larger top metal dimensions and thicker PDMS has been used. The larger dimensions allowed easier handling and alignment of the sensor with the test PCB backplane. The capacitive sensor has been characterized using a commercial readout board (EVAL-AD7746EBZ, Analog Devices Inc.). Figure 5.15a shows the measurement setup where a metal tip attached to a force indenter (MTS Criterion) is slid over the sensor (oriented parallel to the tip) to demonstrate the lateral force application. Figure 5.15b shows the capacitive readout readings, including the extracted values of the normal and shear forces, following the equations in Fig. 5.5.

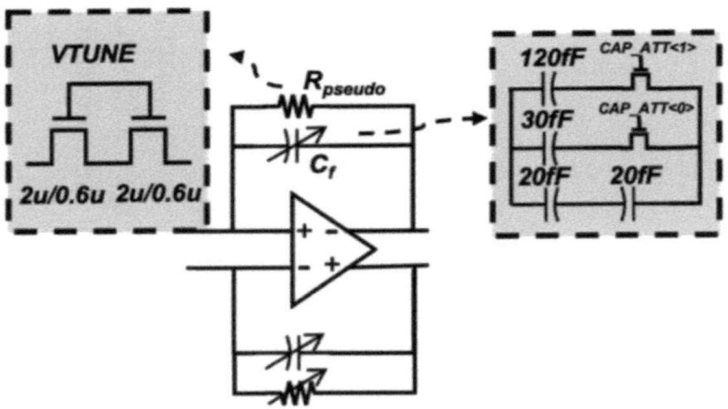

TAXEL X GAIN LOGIC (CAP_ATT<1:0>)		MODE == RAW || FS_X || FN	MODE == FS_Y
GAIN_HI	0	"01" (Cf = 40fF)	"11" (Cf = 160fF)
	1	"00" (Cf = 10fF)	"01" (Cf = 40fF)

TAXEL Y GAIN LOGIC (CAP_ATT<1:0>)		MODE == RAW || FS_Y || FN	MODE == FS_X
GAIN_HI	0	"01" (Cf = 40fF)	"11" (Cf = 160fF)
	1	"00" (Cf = 10fF)	"01" (Cf = 40fF)

Fig. 5.14 Schematic showing how the feedback capacitance changes with both readout mode and gain settings. The schematic also shows the pseudo-resistor sizing

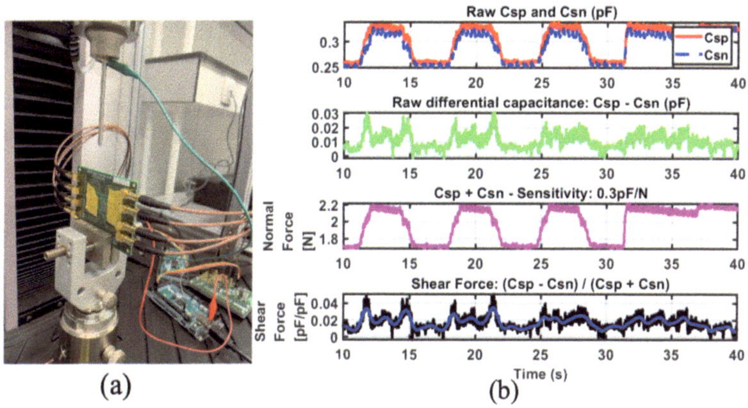

Fig. 5.15 (**a**) Measurement setup and (**b**) capacitive readout readings (using a commercial readout for demonstration) used to detect slip. The two bottom curves indicate the measured normal and shear forces

5.4.2 Proposed Sensor Mismatch Calibration

The mismatch between the two sides of the differential sensor is mainly due to possible misalignments in the top electrode with respect to the bottom electrode. However, this can be calibrated out using extra differential capacitances connected in the $V+$ and $V-$ terminals of the fully differential frontend, C_{attp} and C_{attn}.

The expression for the AFE output is shown in Eq. (5.8). The expression, considering C_{attp} and C_{attn}, becomes:

$$V_{\text{OUT}} = \frac{V_{\text{TX}}\left(C_{\text{attn}}\, C_{\text{sp}} - C_{\text{attp}}\, C_{\text{sn}} - C_f\, C_{\text{sn}} + C_f\, C_{\text{sp}}\right)}{C_f\left(C_{\text{attn}} + C_{\text{attp}} + 2C_f + C_{\text{sp}} + C_{\text{sn}}\right)}. \tag{5.8}$$

Note that the expression is equivalent to Eq. (5.2) if $C_{\text{attp}} = C_{\text{attn}} = 0$.

Static mismatch between C_{sp} and C_{sn} causes the AFE output to deviate from the intended expression of $V_{\text{out}} = 0$ V (differential). To compensate for any sensor mismatch, the calibration procedure is as follows:

1. Measure V_{out} at the idle condition (MODE: raw, no shear and normal forces applied).
2. Assume a starting C_{attp} or C_{attn} value.
3. Set C_{attn} or C_{attp} as a tuning knob until $V_{\text{out}} = 0$, following the following equations:

$$C_{\text{attn}} = \frac{C_{\text{attp}}\, C_{\text{sn}} + C_f\left(C_{\text{sp}} - C_{\text{sn}}\right)}{C_{\text{sp}}}. \tag{5.9}$$

$$C_{\text{attp}} = \frac{C_{\text{attn}}\, C_{\text{sp}} + C_f\left(C_{\text{sn}} - C_{\text{sp}}\right)}{C_{\text{sn}}}. \tag{5.10}$$

From the trim condition $C_f(C_{sp} - C_{sn}) = C_{sp}C_{attn} - C_{sn}C_{attp}$ and bounds $0 \leq C_{attp} \leq C_{attp,\max}$, $0 \leq C_{attn} \leq C_{attn,\max}$, the trimmable mismatch is $-\frac{C_{sn}}{C_f}C_{attp,\max} \leq (C_{sp} - C_{sn}) \leq \frac{C_{sp}}{C_f}C_{attn,\max}$.

Note that, for our testing purposes, the calibration capacitors were put off-chip. However, they can easily be integrated on the chip in future designs.

5.4.3 *Chip Measurements*

The DERMIS design has been prototyped in an NMOS-only 0.6 μm a-IGZO TFT process. The chip measurement setup is shown in Fig. 5.16, while the chip micrograph is shown in Fig. 5.17. For the electrical characterization, the analog frontend (capacitance-to-voltage) output has been measured using external digitally-controlled capacitors as input (1.7–98 pF, with off-chip attenuation capacitance to have the signal fit within the OTA's input common-mode voltage range). The

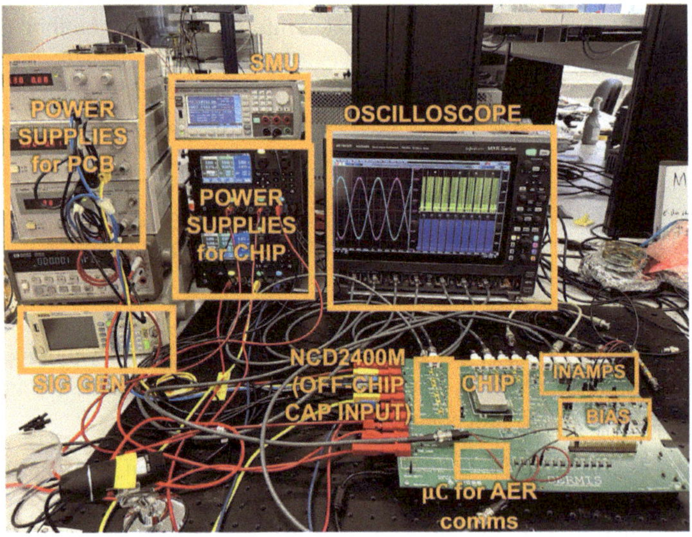

Fig. 5.16 Chip measurement setup. The frontend characterization input signal is applied via a commercial difference amplifier that combines both a DC offset and an AC input signal (from Rigol DG3061A). The replica frontend output and input, and the on-chip temperature sensor output are acquired by a Keysight MXR058A. The chip's current consumption is measured by a Keysight B2061A SMU and Keysight E36312A power supplies. The output spikes during the experiments are recorded off-chip via an ARM Cortex M7 μC (600 MHz clock) with a software-based AER interface communicating with the chip's AER interface

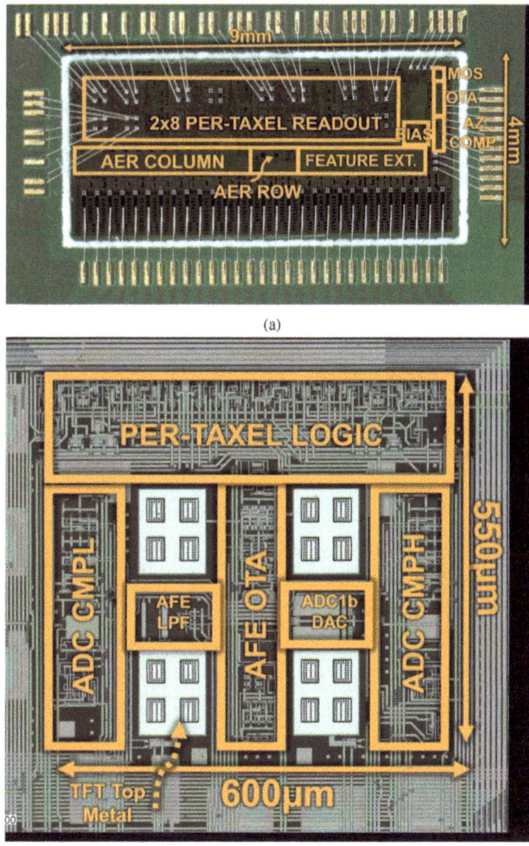

Fig. 5.17 Chip micrograph: (**a**) The DERMIS prototype is shown, highlighting the 2 × 8 taxel array, the AER inteface, the feature extraction logic, and the replica blocks for characterization. The chip's taxels are wirebonded for the off-chip sensor tests. (**b**) The per-taxel readout floorplan, pitch-matched to the 600 µm shear sensing spatial resolution

inherent chopping mechanism (Fig. 5.7) attenuates the flicker noise, resulting in a force-referred input resolution of 2 mN$_{RMS}$ (0.61 fF$_{RMS}$ for 0.3 pF/N sensor sensitivity), which outperforms human skins (Bai et al., 2015). Each per-taxel readout has a power consumption of 72 µW and occupies an area of 0.36 mm^2 (Fig. 5.17b). The per-taxel readout area breakdown is shown in Fig. 5.18. The per-taxel power consumption is dominated by the static power consumption of the digital logic, due to the NMOS-only TFT technology. Similarly, the overall chip power consumption is dominated by the static leakage current of the on-chip AER interface and the feature extraction logic blocks (Fig. 5.19).

The TFT technology used is subject to substantial process variations. The foundry PDK nonetheless does not provide statistical (Monte Carlo) device models. Therefore, mismatch spread simulations were not available, and mismatch variability was assessed from chip-to-chip measurements.

To this end, frontend measurements from four IC samples were collected using identical off-chip capacitors for C_{sp} and C_{sn}, as shown in Fig. 5.20. Three of the four samples exhibit transfer characteristics that follow the predicted frontend output

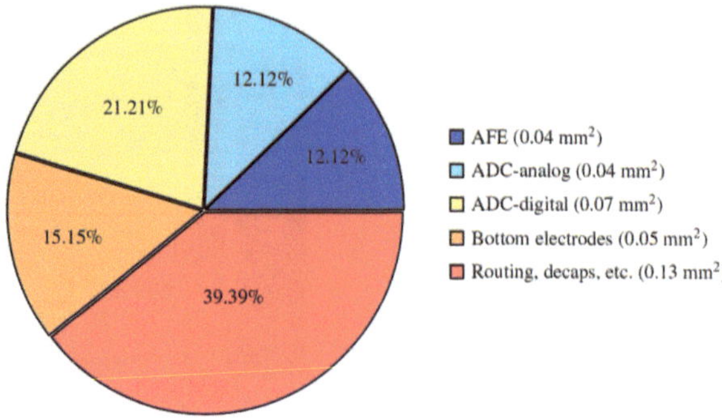

Fig. 5.18 Breakdown of block contributions to the per-taxel readout area. Total readout area is 0.33 mm^2

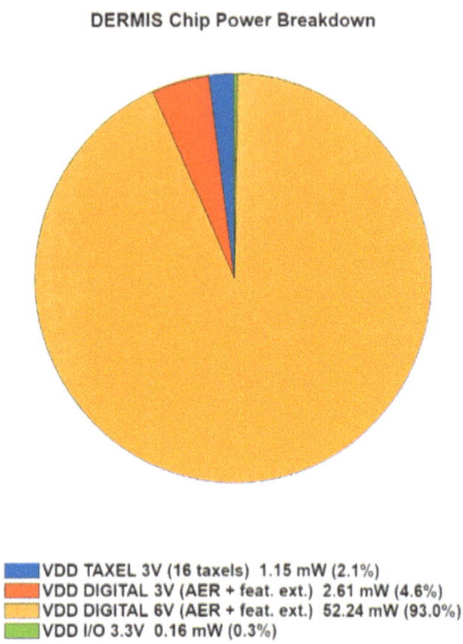

Fig. 5.19 Breakdown of block contributions to the overall chip power consumption

under the assumed finite gain of 36 dB and $C_f = 40$ fF. One sample (Sample 1–Y0) showed a larger offset that shifted the output closer to saturation, consistent with a non-ideal or defect-related deviation. Note that such offset is strongly amplified at the output due to the large $1/\beta$ factor (approximately 43.5) associated with the use of off-chip capacitors. Among the measured devices, Sample 3 most closely matches the ideal output and therefore is taken as being representative of nominal frontend behavior.

The per-taxel level-crossing ADC achieves a peak ENOB of 6.16 *bits* (Fig. 5.21a), resulting in a Walden FOM of 1 nJ/conv·s. Figure 5.21b shows the SNDR versus frequency. It is noted that the SNDR degrades with increasing input frequency, which will be explained further in Sect. 5.5. It is, however, important to note that the target SNDR is still achieved across the target bandwidth of 500 Hz. The simulations in Chap. 3 (Alea et al., 2022) have shown that a low spike-encoding precision (3–5 bits) is not only sufficient, but also prevents overfitting and therefore

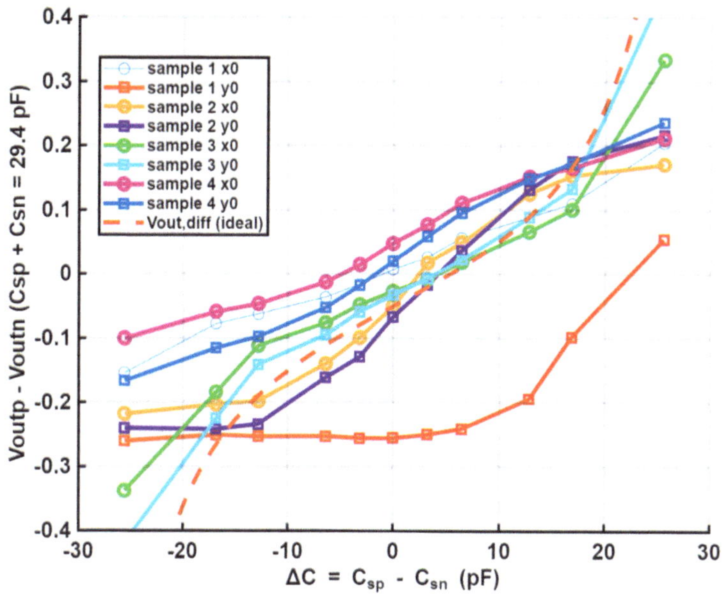

Fig. 5.20 Capacitance-to-voltage frontend output versus the differential capacitance Csp—Csn across multiple samples. The plot shows the ideal voltage output considering a finite gain of 32 dB (measured OTA gain), explaining the slight non-linearity at large $|C_{sp} - C_{sn}|$

maximizing the classification accuracy. This overfitting prevention is likely due to the low spike rates of the low-precision conversion.

Comparisons with the state-of-the-art works in TFT-implemented AFEs and ADCs are shown in Tables 5.4 and 5.5, indicating a superior chip area (400× lower for the AFE; 17× lower for the ADC), power consumption (60× lower), and a comparable Walden FOM. The Walden FOM is approximately twice worse than that of Dandekar and Myny (2025). This is largely because the SAR ADC in Dandekar and Myny (2025) operates at a much higher sampling bandwidth of 32 kHz. TFT-based ADCs suffer when operating at low frequencies because their digital backend draws static power which remains essentially constant. Consequently, because of the lower bandwidth the Walden FOM of our solution degrades.

Although effective in many contexts, applying such a hybrid approach to e-skin readout is less attractive because the additional interconnects required between the TFT and CMOS chips exacerbate the wiring robustness challenges highlighted in Chap. 3. E-skin systems are routinely subjected to direct mechanical contact and repeated deformation, whereas many other hybrid-TFT applications can be fully protected by packaging and do not need to interface mechanically with the environment.

Table 5.6 compares our presented fully-integrated solution to prior state of the art tactile readout systems, highlighting the beyond state of the art 600-μm shear sensing resolution, while being the first to implement a large-area and flexible

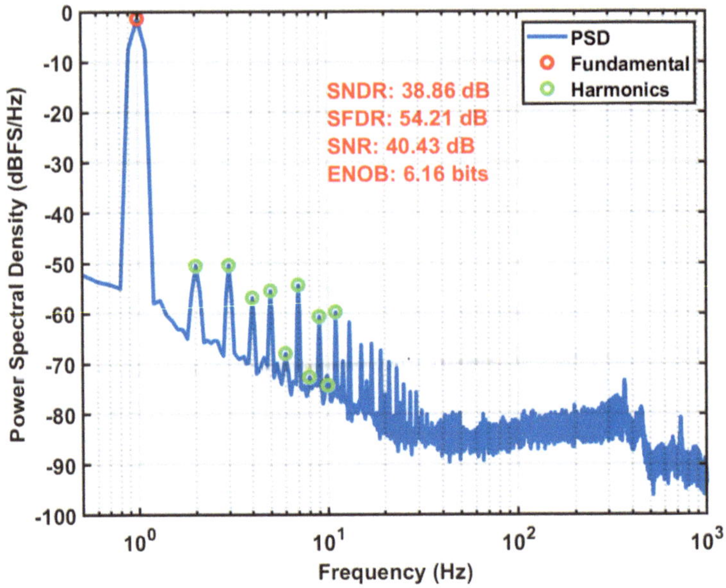

(a)

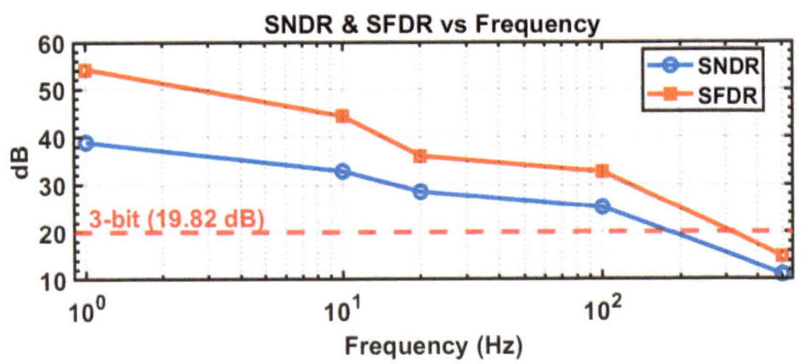

(b)

Fig. 5.21 (**a**) Measured LC-ADC Power Spectral Density (PSD). The input is a 1 Hz sine wave, 6 Vpp (differential). (**b**) Measured SNDR and SFDR as a function of the input frequency. Up to the desired maximum input frequency of 250 Hz, the target precision is achieved

readout system with grasp-state-adaptive force sensing and embedded data compression. Despite being implemented in a coarse-lithography, transistor-performance-constrained TFT process (relative to the Si-CMOS work), the presented work achieves a per-taxel readout area and power consumption comparable to state-of-the-art designs. The 55-nm CMOS design in Chen et al. (2024) occupies only 0.016 mm^2, but this design is not directly comparable to our 0.33-mm^2 implementation in a much coarser 0.6-μm TFT process. Because the area scales roughly with the square of the feature size, the 0.6 μm node inherently requires over two orders of magnitude more area, implying that our design would correspond

Table 5.4 Performance comparison with prior-art TFT-based readout systems

	Moy et al. (2016) ISSCC'16	Genco et al. (2023) JSSC'23	Aygun et al. (2019) ISSCC'19	**This work (DERMIS)**
Application	EEG	EMG	Tactile	**Tactile**
TFT technology	a-Si	a-IGZO	ZnO	**a-IGZO**
Bandwidth (kHz)	0.2	30	N/A[b]	**0.3**
$V_{noise,in}$ (μVRMS)	5.14	94.7	N/A[b]	**6.1**[c]
Area (mm^2)	50	16	60	**0.04**
Power/ch. (mW)	11	0.037	N/A[b]	**0.0006**
V_{DD}(V)	55	26	N/A[b]	**3**
Readout strategy	Hybrid[a]	Hybrid[a]	Hybrid[a]	**TFT-only**

[a]Si-CMOS + TFT
[b]Readout in CMOS
[c]Input-referred, noise gain = $(Cf + Cs)/Cf$
Our implementation in BOLD

Table 5.5 Performance comparison with prior-art TFT-based ADC implementations

	Garripoli et al. (2017) ISSCC'17	Dandekar and Myny (2025) VLSI'25	**DERMIS**
TFT technology	a-IGZO	a-IGZO	**a-IGZO**
μ [cm^2/Vs]/L_{min} [μm]	14 /15	10/0.6	**10/0.6**
ADC architecture	ADSM	SAR	**LC-ADC**
Peak ENOB [bits]	6.35	7.02	**6.16**[a]
Bandwidth [Hz]	300	32,000	**500**
Area (mm^2)	27.9	1.9	**0.11**
Power (mW)	2	1.7	**0.072**
Walden FOM [nJ/c.s.]	39	0.41	**1**

[a]$V_{pp} = 6V$ (full-scale differential), 1 Hz, 1 μs re-timing precision
Our implementation in BOLD

to only about 0.003 mm^2 in 55-nm. When normalized for the technology, our architecture is therefore highly area-efficient and fully competitive with state-of-the-art CMOS designs. Table 5.7 compares the DERMIS capacitance resolution with prior state-of-the-art capacitance-to-digital readouts. The DERMIS work achieves a low capacitance-referred resolution, despite the large flicker noise typical in TFT processes, due to the inherent chopping mechanism.

Compared to the state of the art in commercial slip-sensing tactile sensor systems, this DERMIS work achieves the lowest readout power consumption (36× lower for the taxel array) and the best sensing spatial resolution (3.33× lower), while achieving comparable slip detection latencies (Figs. 5.22 and 5.23).

Table 5.6 Performance comparison of theDERMIS system with recent tactile readout systems

	Chen et al. (2024) JSSC'24	Hao et al. (2024) ISSCC'24	Alea et al. (2023) INTUITIVE	**DERMIS**
Technology	55 nm CMOS	0.18 μm CMOS	0.18 μm CMOS	**0.6 μm a-IGZO TFT**
Sensors	Capacitive	Capacitive	Piezoelectric	**Capacitive**
Shear sensing resolution	N/A[a]	N/A[a]	N/A[a]	**0.6 mm**
Readout method	Event-driven[b]	Frame	Event-driven	**Event-driven**
Readout area (mm^2)[c]	0.016	0.192[d]	0.04	**0.33**
Power (μW)[c]	1.6/7.2	0.75	0.012	**72**
Pitch-matched to sensors	NO	NO	YES	**YES**
Large-area	NO	NO[e]	NO	**YES**
Flexible	NO	NO	NO	**YES**
Grasp-state-adaptive data compression	NO	NO	NO	**YES**

[a]Normal force only
[b]Post-acquisition
[c]Per-taxel
[d]Estimated
[e]Wireless network
Our implementation in BOLD

Table 5.7 Capacitance-to-digital readout comparison

	Gao et al. (2023) ISSCC'23	Jung et al. (2015) ISSCC'15	Oh et al. (2014) VLSI'14	Analog Devices (2005) AD7746	**This work (DERMIS)**
Technology	22 nm CMOS	40 nm CMOS	180 nm CMOS	Commercial IC	**0.6 μm a-IGZO TFT**
Capacitance resolution	37.12 aF_{rms}	0.109% @ 11.3 pF (≈12.3 fF eq.)[a]	≈156 aF_{rms} (from 0–24 pF, 94.7 dB)[b]	4 aF	**0.617 fF** (AFE-limited)[c] **14.35 fF** (ADC-limited)[d]
Architecture	pipelined-SAR	Fully-digital	Incremental ΣΔ + async SAR	ΣΔ CDC	**Direct-input + event-based**

[a]Converted from reported relative error at 11.3 pF.
[b]Derived from reported input range and SNR (assuming full-scale sinusoid).
[c]DERMIS resolution is referred to the input using noise gain: $G = (C_{sp} + C_{sn} + 2C_f + 2C_{att}) / V_{TX}$, with $C_{sp} = C_{sn} = C_{att} = 1.7$ pF (off-chip), $C_f = 40$ fF (on-chip), and $V_{TX} = 3$ V
[d]ADC-limited value from smallest δ-C change that triggers a spike.
Our implementation in BOLD

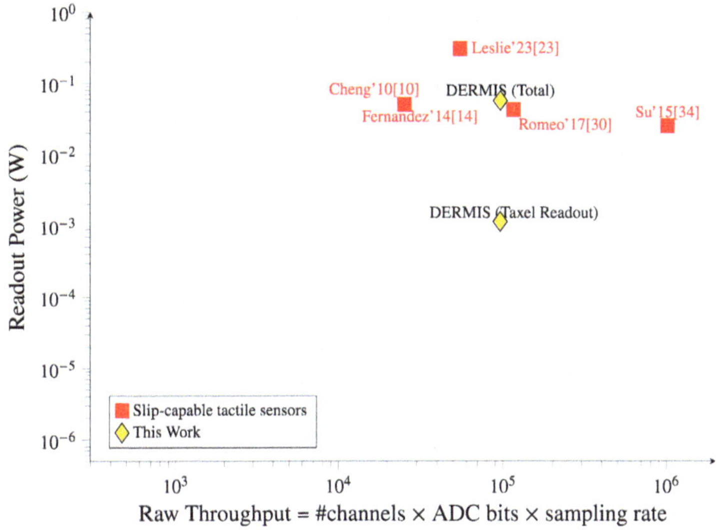

Fig. 5.22 Comparison with existing slip-detecting tactile sensors: Readout power consumption (estimated based on available datasheets) versus the raw throughput

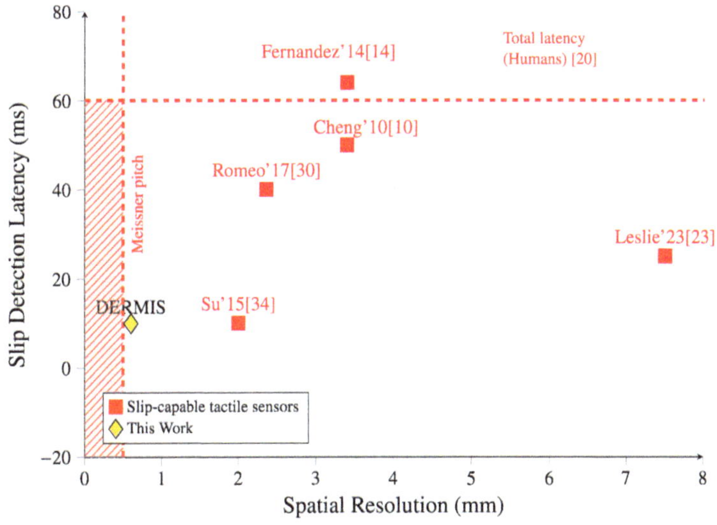

Fig. 5.23 Comparison with existing slip-detecting tactile sensors: the slip detection latency versus the sensing spatial resolution (based on the DERMIS maximum spike rate of 100 $kevent/s/taxel$; not validated in real-life gripping experiments)

5.5 Discussion

In this section, the effect of a correlated noise source to the ADC output SNDR and the dominant sources of the DERMIS chip's power consumption will be discussed. We will also comment on the technology used.

As explained in the previous section, the non-uniform-sampling auto-zeroed readout reduces the effect of the flicker noise on the output spectrum. However, another correlated noise source has been identified as the main culprit to the observed SNDR degradation towards higher input frequencies. Level-crossing ADCs inherently exceed the theoretical limit set by the amplitude quantizer resolution N due to its oversampling properties (Timmermans et al., 2024):

$$\text{SNDR} = 6.02 \cdot N + 1.76[dB] \tag{5.11}$$

Therefore, its SNDR is rather determined by timing errors associated with the samples and the precision of the synchronous re-timing interface (Timmermans et al., 2024) needed to reconstruct the input:

$$\text{SNDR} = -20\log_{10}\left(\delta \times f_{\text{INPUT}}\right) - 11.2[dB] \tag{5.12}$$

where δ is the timing error and f_{INPUT} is the signal input frequency. These timing errors, beyond those intrinsic to the ADC itself, can be compounded by errors introduced during the asynchronous transmission of level-crossing time stamps. Prior LC-ADC implementations (Timmermans et al., 2024; Van Assche & Gielen, 2024) are predominantly single-channel and therefore rely on low-overhead digital logic to transmit samples. In contrast, this work transmits spikes from multiple channels while maintaining compatibility with neuromorphic processing hardware via an address-event representation (AER) interface. An ARM Cortex M7 operating at 600 MHz communicates with the on-chip AER interface using a handshake protocol: it monitors *REQON*/*REQOFF*, reads the spike address in *X* and *Y*, and then asserts an acknowledgment (*ACK*). It has been observed in the prototype chip that unfortunately the on-chip AER interface stalls randomly at intervals correlated to the event rate. The software interface detects a stall (via a timeout period of 100 μs) and resets the on-chip AER block by pulsing *RST* for 100 μs. This results in a period of around 200 μs without any spikes, thus, increasing the signal reconstruction error. Although the precise nature of this stall problem is not yet confirmed, arbiter metastability is a plausible cause. Modeling shows that this behavior injects *colored* noise into the PSD: a corner frequency linked to the event rate degrades the ADC's SNDR (Fig. 5.24). Since the corner frequency rises with the event rate (i.e., larger integrated noise), and the event rate itself increases with input-signal frequency, the SNDR degrades progressively as the input frequency increases. This trend aligns with the measurements and likely explains Fig. 5.25. The signal-dependent corner frequency also rules out flicker noise as the cause of this added colored noise.

Secondly, as shown in Fig. 5.22a, the chip's power consumption is orders of magnitude smaller than prior work, if only the taxel readout array is considered.

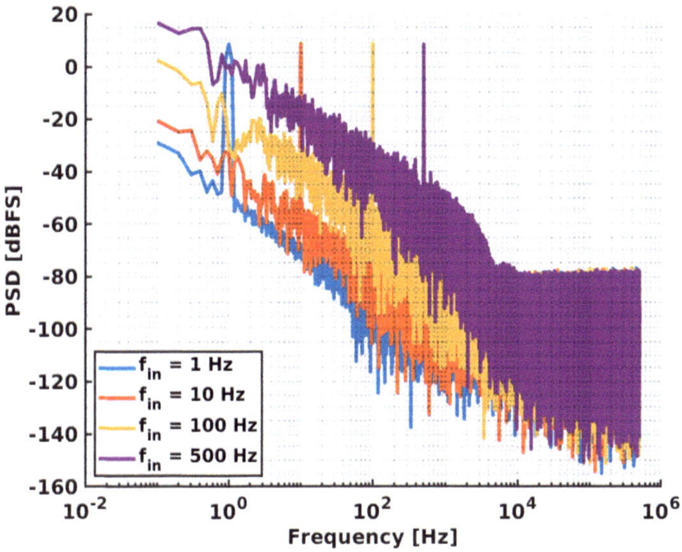

Fig. 5.24 The effect of the observed random AER stalls to the ADC output PSD is modeled. This modeling shows that this behavior injects *colored* noise into the PSD: the corner frequency linked to the event rate degrades the ADC's SNDR for higher input frequencies

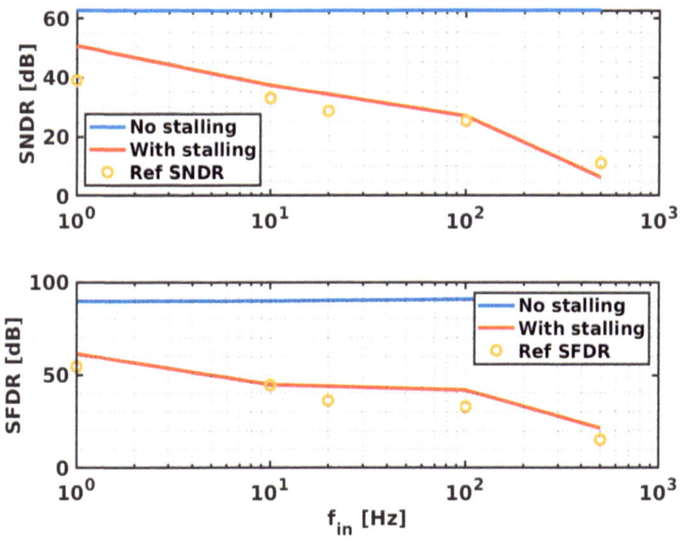

Fig. 5.25 The modeled SNDR as a function of the input frequency with random AER stalls closely resembles that of the measured SNDR

Thus, the chip's system power is dominated by the digital logic in both the AER and the data compression logic. This is due to the large static power consumption observed in the pseudo-CMOS logic cells used in this TFT technology. As discussed in the low-power per-taxel digital block in Sect. 5.3.5, differential logic has a significantly lower static power consumption at the expense of a larger area. In this work, the relative maturity and low area of the available pseudo-CMOS cells has been prioritized over the static power consumption advantage of differential logic cells. The power consumption due to the AER, however, scales only by approximately $log_2(M)$ where M is the number of taxels. The AER logic, which includes the decoders, arbiters, and handshaking logic, grows roughly with the taxel size, so their area and static power overheads follow the same logarithmic scaling.

Finally, this work demonstrates a scalable strategy for realizing electronic skin systems that emulate the high-resolution, energy-efficient tactile sensing of human skin. Although implemented here using thin-film technology, the relative simplicity of the front-end and readout circuitry bodes well for portability and adaptation to other, more exotic, large-area flexible substrates (Kim et al., 2018). The minimalist sensor architecture also enables integrated, high-spatial-resolution shear sensing. The device comprises only a polymer (PDMS) and a metal layer, both of which can be deposited and processed in standard clean-room environments. Consequently, future designs could integrate the sensor during the thin-film transistor (TFT) fabrication step, yielding well-controlled, high-volume, fully integrated tactile-sensor systems.

5.6 Conclusions

This chapter has presented a large step forward toward realizing a low-latency, high-resolution tactile slip sensor system through a fully-integrated end-to-end prototype design in a large-area flexible thin-film technology (TFT). Experimental results of this DERMIS prototype have shown state-of-the-art performance in terms of force resolution, readout area, and power consumption, while integrating biologically-inspired grasp-state-based force sensing and data compression. Leveraging the intrinsic shear- and normal-force detection capabilities of the tactile sensors, the readout system reconfigures based on the current grasp state, achieving the same grasp-state-adaptive force measurements performed in the human somatosensory system. The per-taxel readout achieves a state-of-the-art 72 μW power consumption and occupies an area of 0.36 mm^2 area, while achieving a human-like 2 mN_{RMS} force resolution at 600-μm pitch. This work demonstrates our solution for the first time a fully integrated tactile solution in a *true* large-area prototype of 9×4 mm^2. This DERMIS design, together with the INTUITIVE design of Chap. 4, therefore brings e-skin solutions closer to human tactile capabilities.

References

Alea, M. D., Safa, A., Assche, J. V., & Gielen, G. G. E. (2022). Power-efficient and accurate texture sensing using spiking readouts for high-density e-skins. In *2022 IEEE biomedical circuits and systems conference (BioCAS)* (pp. 359–363).

Alea, M. D., Safa, A., Giacomozzi, F., Adami, A., Temel, I. R., Lorenzelli, L., & Gielen, G. (2023). A fingertip-mimicking 12×16 200μm-resolution e-skin taxel readout chip with per-taxel spiking readout and embedded receptive field processing. In *2023 IEEE symposium on VLSI technology and circuits (VLSI technology and circuits)* (pp. 1–2).

Alea, M. D., Safa, A., Giacomozzi, F., Adami, A., Temel, I. R., Rosa, M. A., Lorenzelli, L., & Gielen, G. (2024). A fingertip-mimicking 12×16 200 μm-resolution e-skin taxel readout chip with per-taxel spiking readout and embedded receptive field processing. *IEEE Transactions on Biomedical Circuits and Systems, 18*(6), 1308–1320.

Analog Devices, Inc. (2005). *AD7745/AD7746: 24-bit capacitance-to-digital converter with temperature sensor* (0th rev ed.). Data Sheet.

Aygun, L. E., Kumar, P., Zheng, Z., Chen, T.-S., Wagner, S., Sturm, J. C., & Verma, N. (2019). 17.3 hybrid system for efficient lae-cmos interfacing in large-scale tactile-sensing skins via tft-based compressed sensing. In *2019 IEEE international solid-state circuits conference - (ISSCC)* (pp. 280–282).

Aygun, L. E., Kumar, P., Zheng, Z., Chen, T.-S., Wagner, S., Sturm, J. C., & Verma, N. (2019). Hybrid lae-cmos force-sensing system employing tft-based compressed sensing for scalability of tactile sensing skins. *IEEE Transactions on Biomedical Circuits and Systems, 13*(6), 1264–1276.

Bai, L., Lehnert, B., Liu, J., Neubarth, N., Dickendesher, T., Nwe, P., Cassidy, C., Woodbury, C., & Ginty, D. (2015). Genetic identification of an expansive mechanoreceptor sensitive to skin stroking. *Cell, 163.*

Boahen, K. (2000). Point-to-point connectivity between neuromorphic chips using address events. *IEEE Transactions on Circuits and Systems II: Analog and Digital Signal Processing, 47*(5), 416–434.

Castaño-Amorós, J., & Gil, P. (2023). Measuring object rotation via visuo-tactile segmentation of grasping region. *IEEE Robotics and Automation Letters, 8*(8), 4537–4544.

Çeliker, H., De Roose, F., Willegems, M., Smout, S., Dehaene, W., & Myny, K. (2024). Analysis and comparison of logic architectures for digital circuits in a-igzo thin-film transistor technologies. *IEEE Journal of Solid-State Circuits, 59*(6), 1858–1870.

Chen, Y., Cai, T., Kuang, Y., Dong, J., Cheng, Z., Zhao, B., & Luo, Y. (2024). A fully dynamic event-driven capacitive sensor interface circuits based on self-reconfigurable sar capacitance-to-digital conversion for high-density robotic tactile sensing. *IEEE Journal of Solid-State Circuits, 59*(8), 2618–2629.

Cheng, M.-Y., Lin, C.-L., & Yang, Y.-J. (2010). Tactile and shear stress sensing array using capacitive mechanisms with floating electrodes. In *2010 IEEE 23rd international conference on micro electro mechanical systems (MEMS)* (pp. 228–231).

Dandekar, M., & Myny, K. (2025). An n-type-only a-igzo thin-film-transistor based nyquist-rate 8-bit cdac+sar adc consuming 1.7mw at 32ksps and achieving 44db sndr. In *2025 symposium on VLSI technology and circuits (VLSI technology and circuits)* (pp. 1–3).

Dandekar, M., Myny, K., & Dehaene, W. (2022). Positive-feedback-based design technique for inherently stable active load toward high-gain amplifiers with unipolar a-igzo tft devices. *IEEE Solid-State Circuits Letters, 5*, 37–40.

Fernández, R., Payo, I., Vazquez, A., & Becedas, J. (2014). Micro-vibration-based slip detection in tactile force sensors. *Sensors, 14*(1), 709–730.

Gao, J., Shen, L., Li, H., Ye, S., Li, J., Xu, X., Cui, J., Gao, Y., Huang, R., & Ye, L. (2023). A 7.9fj/conversion-step and 37.12afrms pipelined-sar capacitance-to-digital converter with kt/c noise cancellation and incomplete-settling-based correlated level shifting. In *IEEE international solid-state circuits conference, ISSCC 2023, San Francisco, CA, USA, February 19–23, 2023* (pp. 346–347). IEEE.

Garripoli, C., van der Steen, J.-L. P. J., Smits, E., Gelinck, G. H., Van Roermund, A. H. M., & Cantatore, E. (2017). 15.3 an a-igzo asynchronous delta-sigma modulator on foil achieving up to 43db snr and 40db sndr in 300hz bandwidth. In *2017 IEEE international solid-state circuits conference (ISSCC)* (pp. 260–261).

Genco, E., Garripoli, C., der Steen, J.-L. P. J. v., Gelinck, G. H., Abdinia, S., Harpe, P., & Cantatore, E. (2023). An emg interface comprising a flexible a-igzo active electrode matrix and a 65-nm cmos ic. *IEEE Journal of Solid-State Circuits, 58*(11), 3138–3149.

Hao, H., Richardson, A. G., Ding, Y., Du, L., Allen, M. G., Spiegel, J. V. D., & Aflatouni, F. (2024). 17.10 a 0.4v, 750nw, individually accessible wireless capacitive sensor interface ic for a tactile sensing network. In *2024 IEEE international solid-state circuits conference (ISSCC)* (Vol. 67, pp. 332–334).

Johansson, R., & Flanagan, J. (2009). Coding and use of tactile signals from the fingertips in object manipulation tasks. *Nature Reviews. Neuroscience, 10*, 345–359.

Jung, W., Jeong, S., Oh, S., Sylvester, D., & Blaauw, D. T. (2015). 27.6 A 0.7pf-to-10nf fully digital capacitance-to-digital converter using iterative delay-chain discharge. In *2015 IEEE international solid-state circuits conference, ISSCC 2015, digest of technical papers, San Francisco, CA, USA, February 22–26, 2015* (pp. 1–3). IEEE.

Kim, Y., Chortos, A., Xu, W., Liu, Y., Oh, J. Y., Son, D., Kang, J., Foudeh, A. M., Zhu, C., Lee, Y., Niu, S., Liu, J., Pfattner, R., Bao, Z., & Lee, T.-W. (2018). A bioinspired flexible organic artificial afferent nerve. *Science, 360*(6392), 998–1003.

Leslie, O., Córdova Bulens, D., & Redmond, S. J. (2023). Design, fabrication, and characterization of a novel optical six-axis distributed force and displacement tactile sensor for dexterous robotic manipulation. *Sensors, 23*, 24.

Liu, F., Deswal, S., Christou, A., Sandamirskaya, Y., Kaboli, M., & Dahiya, R. (2022). Neuro-inspired electronic skin for robots. *Science Robotics, 7*(67), eabl7344.

Moy, T., Huang, L., Rieutort-Louis, W., Wagner, S., Sturm, J. C., & Verma, N. (2016). 16.4 a flexible eeg acquisition and biomarker extraction system based on thin-film electronics. In *2016 IEEE international solid-state circuits conference (ISSCC)* (pp. 294–295).

Oh, S., Jung, W., Yang, K., Blaauw, D. T., & Sylvester, D. (2014). 15.4b incremental sigma-delta capacitance-to-digital converter with zoom-in 9b asynchronous SAR. In *Symposium on VLSI circuits, VLSIC 2014, digest of technical papers, Honolulu, HI, USA, June 10–13, 2014* (pp. 1–2). IEEE.

Park, J.-E., Park, J., Hwang, Y.-H., Oh, J., & Jeong, D.-K. (2019). A noise-immunity-enhanced analog front-end for 36 × 64 touch-screen controllers with 20- V_{pp} noise tolerance at 100 khz. *IEEE Journal of Solid-State Circuits, 54*(5), 1497–1510.

Pragmatic Semiconductor. (2024). *Flexic platform gen 3*. Accessed: 14 Nov 2025.

Romeo, R. A., & Zollo, L. (2020). Methods and sensors for slip detection in robotics: A survey. *IEEE Access, 8*, 73027–73050.

Romeo, R. A., Oddo, C. M., Carrozza, M. C., Guglielmelli, E., & Zollo, L. (2017). Slippage detection with piezoresistive tactile sensors. *Sensors, 17*(8).

Rooijers, T., Huijsing, J. H., & Makinwa, K. A. A. (2022). An auto-zero-stabilized voltage buffer with a quiet chopping scheme and constant sub-pa input current. *IEEE Journal of Solid-State Circuits, 57*(8), 2438–2448.

Rosa, M. B. A., & Kraft, M. (2024). Microengineered flexible pressure sensors with sacrificial molding layer: A novel fabrication approach for improved performance. *Proceedings, 97*(1).

Su, Z., Hausman, K., Chebotar, Y., Molchanov, A., Loeb, G. E., Sukhatme, G. S., & Schaal, S. (2015). Force estimation and slip detection/classification for grip control using a biomimetic tactile sensor. In *2015 IEEE-RAS 15th international conference on humanoid robots (humanoids)* (pp. 297–303).

Timmermans, M., van Oosterhout, K., Fattori, M., Harpe, P., Liu, Y.-H., & Cantatore, E. (2024). A 1.8–65 fj/conv.-step 64-db sndr continuous- time level crossing adc exploiting dynamic self-biasing comparators. *IEEE Journal of Solid-State Circuits, 59*(4), 1194–1203.

Tremblay, M. R., & Cutkosky, M. R. (1993). Estimating friction using incipient slip sensing during a manipulation task. In *Proceedings of the 1993 IEEE international conference on robotics and automation (ICRA), Atlanta, GA, USA, May 1993* (Vol. 1, pp. 429–434).

Tsividis, Y. (2010). Event-driven data acquisition and continuous-time digital signal processing. In *IEEE Custom Integrated Circuits Conference 2010* (pp. 1–8).

Van Assche, J., Carlino, M. F., Alea, M. D., Massaioli, S., & Gielen, G. (2023). From sensor to inference: end-to-end chip design for wearable and implantable biomedical applications. In *2023 IEEE biomedical circuits and systems conference (BioCAS)* (pp. 1–5).

Van Assche, J., & Gielen, G. (2020). Power efficiency comparison of event-driven and fixed-rate signal conversion and compression for biomedical applications. *IEEE Transactions on Biomedical Circuits and Systems, 14*(4), 746–756.

Van Assche, J., & Gielen, G. (2024). Analysis and design of a 10.4-enob 0.92–5.38-μw event-driven level-crossing adc with adaptive clocking for time-sparse edge applications. *IEEE Journal of Solid-State Circuits, 59*(9), 2858–2869.

Zulqarnain, M., & Cantatore, E. (2021). Analog and mixed signal circuit design techniques in flexible unipolar a-igzo tft technology: Challenges and recent trends. *IEEE Open Journal of Circuits and Systems, 2*, 743–756.

Chapter 6
Conclusions and Future Suggestions

This book has focused on the end-to-end design of fully integrated e-skins towards the goal of implementing next-generation tactile sensor systems. Such systems aim to closely approximate the skin's fundamental sensing properties: *multi-modal*, *multi-scale*, and involving *local* processing. We have highlighted our pragmatic, technology-aware approach, drawing on biological inspiration to address the limitations of earlier e-skin research, while maintaining a pick-and-choose strategy that selectively adopts only those biological principles that provide system-level benefits for implementation in artificial electronics substrates. Realizing the full potential of bioinspired designs requires a two-pronged approach of leveraging the integration potentials of VLSI technologies and the careful end-to-end design from the sensors to the readout and processing.

6.1 Summary of the Main Contributions

Introduction and Background Chapter 1 has started with describing how the convergence of bioinspired electronic designs and the proliferation of smart, low-cost sensors have ushered in a new wave of promising biomedical devices with the potential to save and improve lives. Neuroprosthetic users seek to regain human-like touch—restoring functionality, environmental connection, and quality of life. Likewise, advanced robotics would greatly benefit from high-fidelity tactile sensing to achieve more human-like grasping and manipulation. Then, artificial analogs of biological skin—electronic skins—have been described and the challenges involved in achieving the performance and capabilities of their biological counterparts have been described.

After outlining these key challenges, a brief detour was taken into neuroscience to examine the intricacies of the human somatosensory system, aiming to identify fundamental principles that could guide the development of human-like tactile

M. D. Alea, G. Gielen, *Chips That Feel*, Analog Circuits and Signal Processing,
https://doi.org/10.1007/978-3-032-26584-5_6

perception in *e*-skin systems. There, the fundamental properties of human skins were highlighted: its multi-modal, multi-scale properties and its capability to perform local processing. These fundamental properties then formed the guiding principle of this research, resulting in uniting the diverse sensing paradigms of the INTUITIVE and DERMIS projects into a cohesive, functional *e*-skin system.

Literature Review from an End-to-end View Chapter 2 has presented a comprehensive summary of the leading work in e-skin designs. The review has covered state-of-the-art work with a system-level ***end-to-end*** view, ranging from the tactile sensing transducers, the readout and processing electronics, to the data transmission strategies. This end-to-end literature review reflects this text's focus on an end-to-end design of e-skin systems, with the goal of implementing a complete tactile sensor system that approximates the incredible tactile perception capabilities of the human skin. Our literature review has provided the following key insights: (a) the array size scaling of complex force sensors demands sensor miniaturization and integration, (b) scaling up the size of tactile sensor arrays demand aggressive readout constraints, (c) neuromorphic readouts open the door for neuromorphic processing, but the existing designs are insufficient for such scaled designs, and (d) neuromorphic sensing and processing offers system-level processing and data compression benefits, but prior works are limited to software-based implementations only.

Addressing these key limitations of prior e-skin systems, a solution leveraging the high level of integration afforded by VLSI technologies has been proposed. VLSI-based designs, on its own, however, do not necessarily guarantee optimized designs. Thus, an end-to-end coherent design approach, from the sensors to the readout and the processing, needs to be leveraged for obtaining an optimized design, presenting a pathway towards achieving more human-like tactile perception in e-skin systems in terms of their sensing capabilities and their power efficiency.

Two-pronged Approach Towards Next-generation E-skins Chapter 3 has discussed a two-pronged methodology towards bridging the gap between biological tactile sensing and e-skins, in terms of spatial resolution, power efficiency, embedded processing, feature extraction, and data compression. Our pragmatic technology-aware bioinspired approach has then been discussed: following a pick-and-choose approach to adopt biological concepts that optimize e-skin solutions in artificial substrates, rather than biological ones. The approach is to not blindly copy biological concepts but to only borrow techniques that help to address current e-skin bottlenecks. Our key insights were as follows: (a) borrowing the temporally-sparse sampling in neuronal firing using spike-based encoding breaks the trade-off between signal encoding fidelity and power efficiency, especially during periods of sparse tactile activity, and (b) spike-encoding lends well to *self-correcting* SNN-based classifiers, resulting in relaxed readout design constraints, which fit perfectly the goal of (c) true spatiotemporally-sparse sampling, that includes both spatial and temporal compression of sparse tactile inputs.

Prior sparse-sampling neuromorphic approaches perform spike-based conversion in software, thus still using power-inefficient frame-based conversion despite the sparse input. Furthermore, most prior works use off-chip sensors, and have limited sensor count upscaling potential due to wiring, bandwidth, and noise coupling constraints. Therefore, a per-taxel spiking-based readout approach has been presented that results in a power consumption that scales well during periods of inactivity, which are common in tactile applications, while the sensor miniaturization and integration introduce a more sensor-scalable solution. In combination, this allows a true high-density spatiotemporal sampling, as performed in human skin. In this chapter, we have also detailed our breakthroughs towards achieving this goal: (a) sensor miniaturization and integration of both piezoelectric and capacitive shear sensors, with our (b) key takeaways towards implementing a power- and area-efficient per-taxel readout (via event-based level-crossing converters), that fulfills the promise of true spatiotemporal sampling while bridging the gap between biological and artificial e-skin solutions. Our two-pronged approach is highlighted by the complex interplay between design choices spanning from the sensors to the readout and processing circuits (end-to-end design), while leveraging the high levels of integration made possible by current VLSI technologies.

The power consumption modeling and comparison between frame-based and per-taxel spike-based conversions has been derived, showing a clear power-efficiency advantage of event-based spiking readouts over synchronous frame-based approaches, especially at low signal-to-noise ratios. When using a learning-based spiking neural network to classify their output, spike-based readouts have been shown to be robust to readout circuit non-idealities, helping to relax the design constraints on the per-taxel encoding resolution, area, and power consumption. To this end, a neuromorphic level-crossing analog-to-digital converter has been proposed and implemented. By borrowing the adaptive threshold approach in neuronal encoding, and embedding it inside a level-crossing converter architecture, the approach breaks the trade-off between power efficiency and signal encoding fidelity, especially for slow-moving inputs.

INTUITIVE: First Demonstration of Per-taxel E-skin Readout with Integrated Sensors and Local Processing Chapter 4 has presented the first demonstration of an on-chip per-taxel sensor-to-spike encoding *e*-skin solution with embedded complex receptive field (CRF) processing at fingertip-inspired high taxel density, achieving an unprecedented spatial resolution of 200 μm.

This has resulted in the design of the INTUITIVE chip, which has demonstrated the application of our end-to-end design optimizations in conjunction with high levels of integration. The chip features integration of a miniaturized 12×16 polyvinylidene fluoride (PVDF)-based piezoelectric sensor array on a silicon CMOS readout chip with dual per-taxel spiking readout channels. We have shown that the *Spiking Neural Network* (SNN)-based classification performance is relatively *insensitive* to the circuits non-idealities. This allowed to *relax* the readout circuit design targets, in favor of reduced area and power consumption, while fulfilling the tight requirements for per-taxel readout. The experimental results have confirmed that the SNN-based classification of the chip's spatiotemporal spiking output for input tactile stimuli such as texture, flutter, and indentation frequency, achieves excellent accuracies, despite using only a small 256-neuron SNN classifier, a low equivalent spike encoding resolution of 3–5 bits, and a sub-Nyquist population spiking rate. Finally, a comparison of the texture classification accuracies between the two implemented on-chip spike encoder outputs has shown that the proposed neuromorphic level-crossing sampling (N-LCS) architecture with exponentially decaying threshold outperforms the conventional bipolar level-crossing sampling (LCS) architecture with fixed threshold. The prototype chip has been designed and fabricated in 0.18-μm CMOS, and achieves a state-of-the-art power consumption of 12.33 nW per taxel, and 75 μW to 5 mW for the entire chip, depending on the input signals.

The system-level optimizations performed for in the INTUITIVE chip enable near—human-level texture sensing by surpassing the spatial resolution of Merkel cells (0.2 verus 0.4 mm)and reducing the gap in power efficiency by more than 10×.

DERMIS: Extending Per-taxel Readout to Large-area Slip Sensing Chapter 5 then has presented a pathway toward realizing a low-latency, high-resolution tactile slip sensor system through a fully-integrated end-to-end design in a large-area flexible thin-film technology (TFT). A key innovation in this DERMIS work is the miniaturization of the shear and normal force sensors (which enable slip detection), and the design of novel readout circuitry in TFT that reconfigures to independently measure both forces, together with a local processing unit that extracts the relevant grasp-state-dependent parameters from the population spike output.

The DERMIS design also employs our two-pronged approach of end-to-end design and leveraging high levels of integration. A novel floating-electrode differential capacitive sensor, capable of both independent shear and normal force sensing, has been developed and fabricated, allowing for friction-based slip detection algorithms without the need for bulky force/torque sensors. The developed flexible sensor has been optimized for integration on TFT substrates. A novel reconfigurable capacitive tactile sensor frontend then has been designed and implemented using a unipolar 0.6-μm a-IGZO TFT process, enabling the acquisition of context-relevant force parameters depending on the current grasp state (i.e., grasp-state-adaptive). The readout architecture eliminates the need for dedicated charge amplifiers by employing a fully differential frontend with simple switch-phase-based reconfiguration and ratiometric readout, allowing the independent extraction of shear and normal forces as well as the raw differential capacitance. An asynchronous-sampling auto-zero technique, embedded in the level-crossing converter, has been proposed and modeled in this text, resulting in flicker noise reduction without the white noise aliasing common in synchronous-sampling auto-zero techniques.

Besides the use of a TFT technology instead of silicon CMOS, the differential nature of the DERMIS signal conditioning improves upon the INTUITIVE design of Chapter 4: (a) by improving the common-mode noise rejection, and (b) allowing for analog-based arithmetic operations on the differential capacitive sensor, isolating either the normal or shear forces. Leveraging the intrinsic shear- and normal-force detection capabilities of the tactile sensors, the readout system can be reconfigured based on the current grasp state, achieving the same grasp-state-adaptive force measurements as performed in the human somatosensory system. An embedded grasp-state-adaptive feature extraction and data compression logic has been implemented in the unipolar a-IGZO TFT process for the first time, utilizing the population output spikes to detect crucial grasp-state-based parameters such as the onset/offset of contact, the onset/offset of lift-off, initial contact friction, and slip.

The TFT chip has been designed and fabricated in a 0.6 μm a-IGZO process. The experimental results of this DERMIS prototype have shown state-of-the-art performance in terms of force resolution, readout area, and power consumption, while integrating biologically-inspired grasp-state-based force sensing and data compression. The per-taxel readout of the chip consumes a state-of-the-art 72 μW of power consumption, obtains 6.16 ENOB, and occupies an area of 0.36 mm^2, while achieving a human-like 2-mN$_{RMS}$ force resolution (converted from the signal conditioning frontend's input-referred noise). The chip demonstrates our e-skin solution for the first time in a *true* large-area prototype of 9$\times$4 mm^2. It is pitch-matched to a 2$\times$8 differential capacitive sensor array, achieving a state-of-the-art shear and force spatial resolution of 600 μm.

This design brings *e*-skin systems closer to human tactile capabilities. Optimizations in the per-taxel readout of the DERMIS chip have paved the way toward near human-level slip sensing by matching the spatial resolution of Meissner corpuscles, while, for the first time, implementing bioinspired grasp-state-adaptive force sensing which is essential for stable and dexterous manipulation.

Multi-modal and Multi-scale E-skin System: INTUTIVE and DERMIS Combining the high-spatial-resolution, localized texture sensing and complex-receptive-field–based analog processing of the dynamic-force piezoelectric sensors of the INTUITIVE chip with the large-area embedded slip-sensing capabilities of the DERMIS chip—achieved through static and dynamic lateral and shear-force capacitive sensing—advances our vision of a multi-modal, multi-scale *e*-skin system with embedded local processing and with capabilities and performance close to human skin.

6.2 Suggestions for Future Work

We will now describe some suggestions for future work.

6.2.1 Approaching the Power Efficiency of Human Primary Afferents

Our INTUITIVE work explores, for the first time, how learning-based SNN classification can still recover and classify tactile stimuli from spike outputs of imprecise and noisy sensor-to-spike conversions. By significantly scaling down the bias currents, in exchange for reduced noise and bandwidth, and employing simplistic circuit architectures, our work narrowed down the power consumption gap between artifical e-skin and biological firing units to around 5.

Further reducing this gap requires the following:

- significant improvement in the SNN training algorithms to further tolerate even more imprecise, non-linear, and noisier sensor-to-spike conversions;
- significant improvements in the SNN classification algorithms to perform the same classification accuracy despite the inherent loss of higher-frequency information due to the lower-bandwidth amplifiers (due to the reduced DC bias). Current design allows a higher cutoff frequency slightly below 1kHz despite the low bias current by reducing the load capacitances;
- further reduction in the output spike rates to biological levels (50Hz) by more aggressive CRF-based spatial compression (more taxels to compress in one heminode unit);
- simpler charge amplifier architectures at the expense of reduced gain, bandwidth, or linearity (for example, a single non-linear transconductance stage, or a source follower architecture); and
- Increase the use of class-B circuits, which inherently draw no DC current. This is particularly beneficial for the current-mode low-pass filters in the neuromorphic level-crossing design, where currents serve as the main operational variable and continuously consume DC power.

6.2.2 Direct Sensor Integration During CMOS and TFT Fabrication

Our current sensor integration approaches rely on post-processing procedures after the chips have been fabricated (direct deposition on Si-CMOS, alignment post-fabrication on TFT). For the INTUITIVE chip, quadrants of the CMOS wafers are processed during sensor deposition, due to the inherent limitation on the minimum size that can be reliably handled during the process. Future designs could utilize a direct sensor integration on the substrates during the CMOS wafer fabrication itself. The relative simplicity of the sensor structures in both our piezoelectric and capacitive sensor designs lends well to only needing a few extra steps in the wafer fabrication. Furthermore, the sensor yield and precision can be improved by removing the manual alignment procedures post-fabrication. Lastly, the direct integration during wafer fabrication will enable readout designs in smaller CMOS processes, as the handling requirement is already eliminated. Such designs will possibly result in smaller per-taxel area while enabling direct integration of more complex digital logic (e.g., SNN cores).

Currently, the sensor pitch is limited to 200 μm due to constraints of the photolithography process. The implemented per-taxel readout circuits occupy smaller areas, enabling the INTUITIVE chip to integrate both a standard level-crossing converter and a neuromorphic level converter within a single taxel. While wafer-level sensor integration could further enhance spatial resolution, such an approach may be unnecessarily complex and overdesigned for tactile sensing applications.

The minimalistic architecture of the capacitive sensors enables high-spatial-resolution shear sensing. The device comprises only a polymer (PDMS) and a metal layer, both of which can be deposited and processed in standard clean-room environments. Consequently, future designs could integrate the sensor during the thin-film transistor (TFT) fabrication step, yielding well-controlled, high-volume, fully integrated shear-capable tactile-sensor systems.

6.2.3 *Integrating a Machine-learning Accelerator Core*

Our INTUITIVE chip features embedded feature extraction based on complex receptive fields. Such an approach reduces the output data rate by effectively compressing the amount of firing taxels onto a few heminodes. Further output data rate reduction and real-time embedded classification can be achieved by integrating a digital spiking neural network accelerator (SNN) on-chip. Enhancing the power efficiency of the accelerator core though may require porting the INTUITIVE design to more advanced CMOS technology nodes.

To simplify the digital logic, the current DERMIS approach performs simple window-based integration of spike counts to perform the detection of the following parameters: the onset/offset of contact, lift-off, initial friction, and slip. TFT digital logic is inherently power-inefficient in the unipolar TFT technology used in the DERMIS chip, resulting in a large static power consumption. Alternative learning-based approaches to perform detection can be explored on the TFT chip, such as compute-in-memory (Saito et al., 2020) or SNN (Park et al., 2024) cores, making the detection also more robust to the spike output stochasticity.

6.2.4 *Porting the DERMIS Design to Other Large-area Substrates*

Although the a-IGZO TFT process used offers an increased integration due to its fine feature size, relative to other TFT technologies, the relatively higher price per unit area of this technology process may be cost-prohibitive for larger-area requirements (e.g., for demonstrations on robotic arms, body, etc). However, although implemented here using thin-film technology, the relative simplicity and the low-bandwidth operation of the front-end and readout circuitry bode well for portability and adaptation to other, more exotic and cost-effective large-area flexible substrates, such as organic, ZnO, low-temperature polycrystalline silicon (LTPS). The lack of dense digital logic in other substrates can be resolved by putting the

AER logic off-chip, essentially leaving only the shared row and column pull-down lines as the main chip output. Simpler per-taxel digital logic (i.e., a latch and the reset logic) can be implemented with around 10 to 20 custom-made logic gates.

6.2.5 *Circuit Improvements*

The DERMIS TFT chip's system power is dominated by the digital logic in both the AER and the data compression logic. This is due to the large static power consumption observed in the pseudo-CMOS logic cells used in this unipolar technology. An alternative differential logic architecture has a significantly lower static power consumption at the expense of a larger area (roughly twice as large). In this work, the relative maturity (i.e., lower risks) and low area of the available pseudo-CMOS cells have been prioritized over the static power consumption advantage of differential logic cells. This could be altered in future designs. The fully-differential capacitive frontend can tolerate a large common-mode capacitance value, due to the inherent normalization. The main limitation, however, for large input capacitance values is that the input common-mode swing can exceed the OTA's range. A solution used in this book (to allow off-chip sensor inputs) is to put reconfigurable differential-attenuation capacitors at the input terminals. The integration of the off-chip attenuation capacitors will result in a better matching and a reduced noise coupling. The power efficiency of level-crossing ADC designs is fundamentally limited by the power consumption of the continuous-time comparator. Prior techniques, such as adaptive clocking, and adaptive-bias comparators, can be explored to reduce the average power consumption.

6.2.6 *Towards Real-life Demonstrations*

Except for adding the bells and whistles of a standard commercial chip (e.g., design-for-test, serial interface for chip control, better ESD protection, etc), improving the circuit performance and reliability, and improving the sensor yield, commercialization of our approach relies on robust bonding approaches that can be performed massively at scale.

- Improvements in how the sensorized INTUITIVE chip is packaged can be done to improve the e-skin's mechanical robustness while guaranteeing good electrical contacts. Flip-chip approaches on PCBs with a cut-out (to expose the sensorized part) will inherently protect the PCB-to-CMOS connections. Although not implemented, the TFT chip dimension is designed to be wide enough to fit a standard flat flexible cable (FFC). The TFT pitches are intentionally made larger (500 μm) than normal bond pads to accommodate standard FFCs. Bonding can then be done by using anisotropic glue, preventing shorts in-between the pads, with an anisotropic conductive film bonding machine.

- Chip-based sensing approaches enable both localized and distributed sensing by equipping multiple fingers with high-density sensing chips. The amount of physical wiring to route individual fingers, however, can become an engineering challenge. Embedding wireless transceivers at either the chip- or board-level will help minimize the physical wiring at the cost of increased system power consumption, similar to the approach in (Hao et al., 2024). The inherent timed nature of transceiver communications, however, could introduce timing errors to the spike generation. Therefore, to reduce this, local processing (i.e., SNN integration) would be useful.
- Minimizing the number of physical connections or bond pads between the chip and external interfaces improves robustness to mechanical stress, simplifies wiring for robotic hands, and reduces the overall system cost. To achieve this, future fully-integrated e-skin chips should attempt to integrate various blocks such as onboard supply regulators (for both multiple supply generation and noise filtering), internal voltage and current biasing generator (to set DC bias for the circuit blocks), wireless communication transceivers for reduced physical wiring, on-board neuromorphic processing accelerator cores (for local and real-time processing of data, and compression), and a design-for-test bus (for reducing the physical bondpads needed to measure internal nodes), etc.

6.2.7 *Longer-term Vision*

Although the gap between biological and artificial skin has effectively been narrowed down by our DERMIS and INTUITIVE approaches in terms of power efficiency and local processing of high-spatial-resolution texture sensing and large-area slip sensing, there are still many other features of biological skin that remain unaddressed, such as growth, self-healing, and lifelong adaptation of the somatosensory neural processing.

6.2.7.1 Self-healing Electronics

Real biological skin, when exposed to damage, tries to repair this damage. Obviously, this is not the approach in existing electronic systems. Small damages result in a degradation in performance, while a complete breakdown renders the system invalid. Future systems may adopt this self-healing nature to reduce the maintenance and cost, and increase the long-term reliability of future e-skin designs. Current work in self-healing electronic skins (Chen et al., 2022; Son et al., 2018) has observed that polymers that form the soft substrate, the sensor layer, and the conductive tracks exhibit such self-healing effect. This has the potential to restore both the mechanical integrity and the electrical functions of future e-skin systems.

Obviously, current electronic skin systems are orders of magnitude more complex than current self-healing systems. Adapting this approach would then require to implement the complete and complex electronics on a polymer substrate similar to the one in Chen et al. (2022), or altenatively to develop self-healing approaches in both rigid silicon-CMOS or flexible TFT substrates.

For tactile systems, an interesting twist is to couple these material advances with system- and circuit-level *self-reconfiguration*. The electronics can periodically run built-in self-tests to detect damaged regions, and execute in parallel, the physical repair of the damaged area while temporarily re-configuring the system such that it bypasses the damaged region without severe degradation in its overall performance or functionality. Obviously, this has applications beyond just electronic skin systems. Therefore, such self-healing electronics approaches will undoubtedly gain more interest in the scientific community.

6.2.7.2 Biohybrid Skin

On the opposite end, one can imagine totally moving the implementation of electronic skin from artificial substrates to biological ones through lab-grown skin organoids (Sun et al., 2021). Such an approach will approximate better the skin's compliance and the mechanoreceptors' sensitivity to small changes in the complex forces. Sun et al. (2021) suggests that mechanoreceptor- and primary-afferent-rich lab-grown skin can be grown *in vitro*, mirroring the somatosensory sensing and neural pathways in the skin. Upstream processing can still be retained to using conventional electronics, as pioneering works show that lab-grown brain organoids and other biological three-dimensional neural networks can be interfaced with high-density bioelectronics (Wang et al., 2025; Faravelli et al., 2025).

In the future, *biohybrid e-skin* systems can still with interfaced by upstream conventional electronics (for compatibility with electronic computing systems). Instead of interfacing single mechanoreceptors, bundles of primary afferents can be read out and stimulated, much like a peripheral nerve interface, effectively offloading the local processing capabilities to the lab-grown biological substrate.

References

Chen, J., Wang, L., Xu, X., Liu, G., Liu, H., Qiao, Y., Chen, J., Cao, S., Cha, Q., & Wang, T. (2022). Self-healing materials-based electronic skin: Mechanism, development and applications. *Gels, 8*(6), 356.

Faravelli, I., Antón-Bolaños, N., Brown, J. R., & Arlotta, P. (2025). Brain organoids: Tools for understanding the uniqueness and individual variability of the human brain. *Annual Review of Genomics and Human Genetics, 26*, 299–320.

Hao, H., Richardson, A. G., Ding, Y., Du, L., Allen, M. G., Spiegel, J. V. D., & Aflatouni, F. (2024). 17.10 a 0.4v, 750nw, individually accessible wireless capacitive sensor interface ic for a tactile sensing network. In *2024 IEEE international solid-state circuits conference (ISSCC)* (Vol. 67, pp. 332–334).

Park, J., Yun, Y., Kim, M., & Lee, S.-Y. (2024). Simulation and optimization of igzo-based neuromorphic system for spiking neural networks. *IEEE Journal of the Electron Devices Society, 12*, 228–235.

Saito, D., Doevenspeck, J., Cosemans, S., Oh, H., Perumkunnil, M., Papistas, I. A., Belmonte, A., Rassoul, N., Delhougne, R., Kar, G., Debacker, P., Mallik, A., Verkest, D., & Na, M. H. (2020). Igzo-based compute cell for analog in-memory computing—dtco analysis to enable ultralow-power ai at edge. *IEEE Transactions on Electron Devices, 67*(11), 4616–4620.

Son, D., Kang, J., Vardoulis, O., Kim, Y., Matsuhisa, N., Oh, J. Y., To, J. W. F., Mun, J., Katsumata, T., Liu, Y., McGuire, A. F., Krason, M., Molina-Lopez, F., Ham, J., Kraft, U., Lee, Y., Yun, Y., Tok, J. B. H., & Bao, Z. (2018). An integrated self-healable electronic skin system fabricated via dynamic reconstruction of a nanostructured conducting network. *Nature Nanotechnology, 13*(11), 1057–1065.

Sun, H., Zhang, Y.-X., & Li, Y.-M. (2021). Generation of skin organoids: Potential opportunities and challenges. *Frontiers in Cell and Developmental Biology, 9*, 709824.

Wang, Q., Dong, X., Jiang, D., Tian, S., Qiu, Y., Zhu, Y., Wu, J., Shang, S., Zhang, Y., Wang, P., & Zhuang, L. (2025). Bioelectronic interfaces and sensors for neural organoids. *Microsystems & Nanoengineering, 11*(1), 172.

Appendix A
Estimating the Power Consumption of Afferent Firing

A.1 Energy Estimate of Aβ Tactile Afferents

Both Merkel (SA-I) and Meissner (FA-I) mechanoreceptors project through thick, myelinated Aβ primary afferents. We approximate the energy budget derived by Harris and Attwell (2012) for CNS myelinated axons. The following is a rough estimate of the power consumption (estimated for rodents), as there is no equivalent estimation for human tactile afferents, as far as the author knows.

For a myelinated axon, the ATP molecules needed *per spike* for one node are given by Harris and Attwell (in eq. 9 of their paper) as

$$N_{\text{ATP,node}} = \frac{\left(2\pi r C_A L_{\text{node}} + C_L L_{\text{internode}}\right) \Delta V \, f_{\text{Na}}}{3e}, \tag{A.1}$$

where

r is the axon radius (inside myelin), here $r = 0.89/2\,\mu\text{m} = 0.445 \times 10^{-6}\,\text{m}$,
C_A is the nodal membrane capacitance per unit area, $C_A = 10^{-2}\,\text{F/m}^2$,
L_{node} is the node length, $L_{\text{node}} = 0.8\,\mu\text{m} = 0.8 \times 10^{-6}\,\text{m}$,
C_L is the internodal capacitance per unit length, $C_L = 2.06 \times 10^{-9}\,\text{F/m}$,
$L_{\text{internode}}$ is the internode length, $L_{\text{internode}} = 240\,\mu\text{m} = 240 \times 10^{-6}\,\text{m}$,
ΔV is the action potential amplitude, $\Delta V = 0.1\,\text{V}$,
f_{Na} is the Na^+ overlap factor, $f_{\text{Na}} = 1.3$ (Alle et al., 2009; Carter & Bean, 2009),
e is the elementary charge, $e = 1.602 \times 10^{-19}\,\text{C}$.

First compute the effective capacitance of one segment:

$$C_{\text{node}} = 2\pi r C_A L_{\text{node}}, \tag{A.2}$$

$$C_{\text{internode}} = C_L L_{\text{internode}}, \tag{A.3}$$

$$C_{\text{seg}} = C_{\text{node}} + C_{\text{internode}}. \tag{A.4}$$

M. D. Alea, G. Gielen, *Chips That Feel*, Analog Circuits and Signal Processing,
https://doi.org/10.1007/978-3-032-26584-5

Numerically,

$$C_{\text{node}} = 2\pi(0.445 \times 10^{-6})(10^{-2})(0.8 \times 10^{-6})\,\text{F} \tag{A.5}$$
$$\approx 2.24 \times 10^{-14}\,\text{F}, \tag{A.6}$$

$$C_{\text{internode}} = (2.06 \times 10^{-9})(240 \times 10^{-6})\,\text{F} \tag{A.7}$$
$$\approx 4.94 \times 10^{-13}\,\text{F}, \tag{A.8}$$

$$C_{\text{seg}} \approx (2.24 \times 10^{-14} + 4.94 \times 10^{-13})\,\text{F} \tag{A.9}$$
$$\approx 5.17 \times 10^{-13}\,\text{F}. \tag{A.10}$$

The ATP molecules used per spike for one segment are then:

$$N_{\text{ATP,node}} = \frac{C_{\text{seg}}\,\Delta V\, f_{\text{Na}}}{3e} \tag{A.11}$$
$$= \frac{(5.17 \times 10^{-13}\,\text{F})(0.1\,\text{V})(1.3)}{3 \times 1.602 \times 10^{-19}\,\text{C}} \tag{A.12}$$
$$\approx 1.4 \times 10^{5}\ \text{ATP per spike per segment.} \tag{A.13}$$

The number of such segments per meter is approximately (from Table 1 of Harris & Attwell, 2012):

$$N_{\text{seg/m}} \approx \frac{1}{L_{\text{internode}}} = \frac{1}{240 \times 10^{-6}\,\text{m}} \approx 4.17 \times 10^{3}\ \text{segments/m}. \tag{A.14}$$

Thus, the ATP molecules consumed per spike per meter of axon are:

$$N_{\text{ATP/m}} = N_{\text{ATP,node}}\, N_{\text{seg/m}} \tag{A.15}$$
$$\approx (1.4 \times 10^{5}) \times (4.17 \times 10^{3}) \tag{A.16}$$
$$\approx 5.8 \times 10^{8}\ \text{ATP/(spike} \cdot m). \tag{A.17}$$

To convert this to energy, we take the Gibbs free energy per ATP molecule (Berg et al., 1995) as:

$$\Delta G_{\text{ATP}} \approx 50\,000\,\text{J/mol}, \tag{A.18}$$
$$N_A = 6.022 \times 10^{23}\,\text{mol}^{-1}, \tag{A.19}$$
$$\Rightarrow \varepsilon_{\text{ATP}} = \frac{\Delta G_{\text{ATP}}}{N_A} \approx \frac{50\,000}{6.022 \times 10^{23}}\,\text{J} \tag{A.20}$$
$$\approx 8.3 \times 10^{-20}\,\text{J/ATP}. \tag{A.21}$$

The energy per spike per meter is then:

$$E_{\text{spike/m}} = N_{\text{ATP/m}}\, \varepsilon_{\text{ATP}} \tag{A.22}$$

$$\approx (5.8 \times 10^8) \times (8.3 \times 10^{-20})\,\text{J} \tag{A.23}$$

$$\approx 4.8 \times 10^{-11}\,\text{J/m} \tag{A.24}$$

$$= 48\,\text{pJ/m per spike.} \tag{A.25}$$

Finally, the mean power for a myelinated Aβ afferent of length $L = 1$ m firing at $f = 50$ Hz is:

$$P = E_{\text{spike/m}} \cdot L \cdot f \tag{A.26}$$

$$= (48 \times 10^{-12}\,\text{J/m}) \times (1\,\text{m}) \times (50\,\text{s}^{-1}) \tag{A.27}$$

$$= 2.4 \times 10^{-9}\,\text{W} \tag{A.28}$$

$$= 2.4\,\text{nW.} \tag{A.29}$$

Appendix B
Estimating the Maximum Force Allowed by the Sensor

The following estimates are derived from the proposed DERMIS sensor dimensions, assuming a PDMS dielectric with thickness of 125 μm.

B.1 Maximum Shear Load Estimate

Given a bonded PDMS footprint $A = (1.4\,\text{mm})(7.5\,\text{mm}) = 1.05 \times 10^{-5}\ \text{m}^2$, thickness $t = 125\,\mu\text{m}$, and maximum lateral displacement $\Delta x_{\text{max}} = 50\,\mu\text{m}$, the simple-shear strain is

$$\gamma_{\text{max}} = \frac{\Delta x_{\text{max}}}{t} = \frac{50}{125} = 0.40. \tag{B.1}$$

Assuming linear isotropic elasticity, $\tau = G\gamma$ with

$$G = \frac{E}{2(1+\nu)} \approx \frac{E}{3} \quad (\nu \approx 0.5), \tag{B.2}$$

and using $E = 4.42$ MPa for Sylgard 184 cured at 150°C (Moučka et al., 2021),

$$\tau_{\text{max}} = G\gamma_{\text{max}} \approx \left(\frac{4.42 \times 10^6}{3}\right)(0.40) = 5.9 \times 10^5\ \text{Pa}. \tag{B.3}$$

The corresponding maximum shear force over the footprint is:

$$F_{\text{max}} = \tau_{\text{max}} A = (5.9 \times 10^5)(1.05 \times 10^{-5}) = 6.2\,\text{N}. \tag{B.4}$$

M. D. Alea, G. Gielen, *Chips That Feel*, Analog Circuits and Signal Processing,
https://doi.org/10.1007/978-3-032-26584-5

B.2 Maximum Normal (Compressive) Load Estimate

For a PDMS layer of thickness $t = 125\,\mu\text{m}$ and a loaded footprint $A = (1.4\,\text{mm})(7.5\,\text{mm}) = 1.05 \times 10^{-5}\ \text{m}^2$, an imposed normal indentation Δz gives the following nominal compressive strain:

$$\varepsilon = \frac{\Delta z}{t}. \tag{B.5}$$

Assuming first-order uniaxial compression, the nominal compressive stress is:

$$\sigma = E\,\varepsilon, \tag{B.6}$$

and the corresponding normal force is:

$$F_N = \sigma A = E\left(\frac{\Delta z}{t}\right) A. \tag{B.7}$$

Using $E = 4.42\,\text{MPa}$ for Sylgard 184 cured at 150° C (Moučka et al., 2021), the force scales as $F_N \approx 3.71 \times 10^7$, Δz (with Δz in meters). For example, $\Delta z = 50\,\mu\text{m}$, then $F_N = 18.6\,\text{N}$, while if $\Delta z = 100\,\mu\text{m}$, then $F_N = 37.1\,\text{N}$.

References

Alle, H., Roth, A., & Geiger, J. R. P. (2009). Energy-efficient action potentials in hippocampal mossy fibers. *Science, 325*(5946), 1405–1408.

Berg, J. M., Tymoczko, J. L., & Stryer, L. (1995). *Biochemistry* (4th ed.). W. H. Freeman.

Carter, B. C., & Bean, B. P. (2009). Sodium entry during action potentials of mammalian neurons: incomplete inactivation and reduced metabolic efficiency in fast-spiking neurons. *Neuron, 64*(6), 898–909.

Harris, J. J., & Attwell, D. (2012). The energetics of cns white matter. *Journal of Neuroscience, 32*(1), 356–371.

Moučka, R., Sedlačík, M., Osička, J., & Pata, V. (2021). Mechanical properties of bulk sylgard 184 and its extension with silicone oil. *Scientific Reports, 11*, 19090.

MIX
Papier aus verantwortungsvollen Quellen
Paper from responsible sources
FSC® C105338

If you have any concerns about our products,
you can contact us on
ProductSafety@springernature.com

In case Publisher is established outside the EU,
the EU authorized representative is:
Springer Nature Customer Service Center GmbH
Europaplatz 3, 69115 Heidelberg, Germany

Printed by Libri Plureos GmbH
in Hamburg, Germany